ADVANCE PRAISE FOR
The Edge of Knowledge

"Krauss has a genius for making science understandable and exciting in equal parts. This account of some of the key puzzles in physics is a wonderfully lucid and stimulating introduction, which only a real scientist with an enthusiastic gift for communication could achieve. Krauss ticks both boxes handsomely. It adds to his remarkable body of work in making the world more scientifically literate and greatly more rational."

—**Anthony Grayling**, Professor of Philosophy and principal, New College of Humanities, and author, most recently of *For the Good of the World*

"Lawrence Krauss knows science. Even more impressive, Lawrence can explain science (even to me). This book is about the crucial stuff that Lawrence knows and some stuff that even Lawrence doesn't know (and neither does anyone else). He explains clearly what no one knows, so we can all help figure it out."

—**Penn Jillette**, magician, actor, TV host, and author, most recently of *Random*

"In *The Edge of Knowledge*, Lawrence Krauss dispels the classroom notion of science as a collection of facts, showing that it is really a disciplined way of exploring the unknown. His deft traverse from the vastness of the universe to the intricacies of life lays bare the enduring unknowns that will motivate research for years to come."

—**Andrew Knoll**, Fisher Research Professor of Natural History and Research Professor of Earth and Planetary Sciences, Harvard University, and author, most recently, of *A Brief History of Earth*

"This fascinating book offers a highly readable exposition of the fundamental questions that perplex us most—those that lie on (or beyond) the current frontiers of human knowledge. Lawrence Krauss is a fine scientist with a broad perspective. And he's also an excellent writer, able to expound deep mysteries in language that's always clear, and often entertaining too. Anyone with an enquiring mind should enjoy this book—it deserves very wide readership."

—**Lord Martin Rees**, Astronomer Royal, former President of the Royal Society, and Master of Trinity College, Cambridge

Also by Lawrence M. Krauss

The Physics of Climate Change
The Greatest Story Ever Told—So Far
A Universe from Nothing
Quantum Man
Hiding in the Mirror
Atom
Beyond Star Trek
The Physics of Star Trek
Quintessence
Fear of Physics
The Fifth Essence

THE
EDGE
OF
KNOWLEDGE

THE
EDGE
OF
KNOWLEDGE

UNSOLVED MYSTERIES
OF THE COSMOS

LAWRENCE M. KRAUSS

Post Hill
PRESS

A POST HILL PRESS BOOK

The Edge of Knowledge:
Unsolved Mysteries of the Cosmos
© 2023 by Lawrence M. Krauss
All Rights Reserved

ISBN: 978-1-63758-856-7
ISBN (eBook): 978-1-63758-857-4

Cover design by Hampton Lamoureux
Interior design and composition by Greg Johnson, Textbook Perfect

Post Hill Press
New York • Nashville
posthillpress.com

Published in the United States of America
1 2 3 4 5 6 7 8 9 10

To my family and my real friends,
for your generous support through turbulent times.

*I don't feel frightened not knowing things, by being
lost in a mysterious universe without any purpose,
which is the way it really is as far as I can tell.*

—RICHARD P. FEYNMAN

*As we know, there are known knowns;
there are things we know we know. We also know
there are known unknowns; that is to say we know
there are some things we do not know.
But there are also unknown unknowns—
the ones we don't know we don't know.*

—DONALD RUMSFELD

CONTENTS

FOREWORD

Three of the most important words in science are: "I don't know." Therein lies the beginning of enlightenment because not knowing implies a universe of opportunities—the possibility of discovery and of surprise.

If history is any guide, there is a lot more about the universe that we don't know than we do. Sometimes that is incorrectly taken to imply that we know almost nothing. In truth, we know a great deal, and that guides us in our search to learn more. But the recognition that many cosmic mysteries still remain provides a long-term hopefulness to the scientific enterprise, not to mention a kind of cosmic job security.

The limits of the world we understand have moved further and further beyond the universe of our direct experience over the past five hundred years of modern science. Yet the fundamental mysteries of existence persist: How did our universe begin, if it even had a beginning? How will it end? How big is it? What lies beyond what we can see? What are the fundamental laws governing our existence? Are those laws the same everywhere? What is the world of our experience made of? What remains hidden? How did life on earth arise? Are we alone? What is consciousness? Is human consciousness unique?

These questions continue to drive our explorations, but like venturing further into dark woods, persistent cosmic mysteries get deeper and more exciting. It is a lesson of history that each discovery raises new puzzles, giving us a vital new perspective on what the significance and meaning of these same fundamental questions are.

The mysteries are moving targets, and they define the scientific forefront—the threshold of the unknown. To explore that threshold is to gain a deeper understanding of just how far science has progressed. That is the purpose of this book.

Appreciating exactly what we don't know requires some intellectual baggage, which can be extensive for those preparing for a career in science. Getting to that point represents the transition between being a student and becoming a professional researcher. But to get a basic perspective, rather than mastery, of the current limits of knowledge is something that can be done more readily.

This book is designed to give readers such a perspective. It is structured around the big outstanding mysteries mentioned earlier, with chapters broadly based on time, space, matter, life, and consciousness. Each section addresses a series of apparent mysteries that will be listed at the beginning of that section.

The result, I hope, will be a celebration of knowledge rather than of ignorance. It is an invitation to ponder and appreciate the universe in which we live.

* * *

The physicist Richard Feynman speculated about whether we might ultimately develop a single theory that explained all cosmic phenomena. He doubted it, and so do I. But as he put it, even if reality is like some infinite cosmic onion and each new development is merely another layer peeled back, that would be good enough for him. He just wanted to know more about the universe today than he did yesterday. And when he did, I suspect he was surprised. Because that, to me, is the most fascinating aspect of the cosmos: that it keeps surprising us. The imagination of nature is far greater than the imagination of humans. In my own work, every day I am surprised if I am not surprised.

That is why we have to keep probing with our experiments. If we merely theorized or speculated, it is most likely we would wander off on the wrong trail. Our experiments keep us on the right track, and they keep us honest. We try to follow a path laid out by nature, but the markers are hidden in advance, and the destination is not always clear.

Donald Rumsfeld's infamous quote resonates here. The most exciting discoveries in science generally involve the "unknown unknowns" because that is where the greatest surprises lie, and where new trajectories of knowledge begin.

But if we knew what these unknown unknowns were, they wouldn't *be* unknown unknowns. So, as we ponder nature at the limits of our knowledge, and perhaps also at the limits of our imagination, we make do with what we have, the known unknowns. Fortunately, by pushing on these, we are often repaid with unexpected answers, and new questions.

It is my fervent hope that in a generation from now, many of the universe's mysteries as I have framed them here will seem quaint or poorly conceived. The questions may remain unchanged, but our perspective of them will likely be completely altered. This book then may serve for that future generation as a reminder of how much science has progressed, much as Sir James Jeans's classic 1930 masterpiece, *The Mysterious Universe*—which had a huge impact on the public's perception of science at that time—does for us now nearly a century after its publication.

I hope to be around to watch that happen.

* * *

This book wouldn't have been written without the gentle prodding of my wonderful British publisher Anthony Cheetham, who threw out a challenge I couldn't resist accepting. Throughout the process, he and his colleagues at Head of Zeus publishing, now publishers of two of my books, were patient and persistent. I thank them for both of these attributes and hope that this final product does justice to their hopes.

My US publisher, Adam Bellow at Post Hill Press (who published my last book as well), has provided much added wisdom, and his colleagues, including my wonderful copyeditor, Andrew Horgan, have worked hard at every stage to improve the manuscript.

While many of the areas I discuss here overlap with my own expertise as a physicist, I have benefitted greatly from discussions with and books by a number of colleagues from other disciplines over the years who have shaped my understanding, stimulated my thinking, and often exposed my own misconceptions. This list includes Andrew Knoll, Noam Chomsky, Richard Dawkins, Joseph LeDoux, Steven Pinker, Ian McEwan, George Church, Nancy Dahl, John Sutherland, and Jonathan Rauch, among others. In addition, Noam Chomsky, Richard Dawkins, Jana Lenzova, and Neil deGrasse Tyson kindly read early drafts of the manuscript and made many valuable suggestions that I have incorporated in the final version. I thank them for their generosity, friendship, and wisdom. Finally, I want to thank my many physics colleagues and students over the years for stimulating, changing, and often correcting my own understanding, and for helping guide my interests and motivating me to continue to explore the universe's known unknowns.

1

<u>TIME</u>

Is time universal?
Does time have a beginning?
Can time end?
Is time travel possible?

Lost time is never found again.

—BENJAMIN FRANKLIN

Here we are, Mr. Pilgrim, trapped in the amber of this moment. There is no why.

—KURT VONNEGUT

It's being here now that's important. There's no past and there's no future. Time is a very misleading thing. All there is ever, is the now. We can gain experience from the past, but we can't relive it; and we can hope for the future, but we don't know if there is one.

—GEORGE HARRISON

Time is the most personal feature of our existence. It is the thread that holds the drama of our lives together, and it is at the center of all literature, good and bad. It feeds the heart of tragedy and the pulse of adventure. Yet it remains so mysterious that some seriously question whether it even really exists.

Einstein once joked while explaining relativity that when talking to someone enchanting, an hour can seem like a minute, while if sitting on a hot stove, a minute can seem like an hour. Although he was being facetious, there is an important element of truth here. The perception of passing time depends on your state of mind—whether you are bored or excited, for example.

Regardless of your state of mind, time is precious. Due to modern medicine, most of us outlive the traditional three score and ten years of the Bible, but our time on earth is nevertheless finite. We don't get a repeat, to paraphrase Benjamin Franklin, and anyone who has seen a particularly bad movie knows that there is no recompense for the lost hours.

Much philosophical ink has been spilled about the question of whether physics implies that time itself is fundamental or an illusion. While I will briefly touch on this question later, I think that like many philosophical discussions, it misses the key issues that physicists, and everyone else for that matter, actually worry about. The fact that time governs almost every aspect of our daily lives is undeniable. The suggestion that time may be an illusion doesn't do much for someone who rushes onto a train platform only to find that their 5:50 commuter train has just left.

As it happens, it was a consideration of trains that prompted Albert Einstein to change the notion of time as a physical quantity.

Up to that point, the effort to synchronize clocks across long distances had been a major challenge, especially given that international commerce and warfare were primarily carried out at sea. An accurate determination of longitude while traveling along an east-west axis on the ocean is essential in order to know one's position relative to one's destination.

This became possible when the local time (determined by the position of the sun) could be compared to the time at one's origin, which required a clock that could remain accurate for long naval voyages. The problem of measuring longitude was deemed of such importance in Britain that in 1714, Parliament created a public reward of ten thousand to twenty thousand pounds for generating a method to do so, with the amount depending upon the accuracy of longitude obtained. In 1730, John Harrison, a carpenter and clockmaker, presented his design for a marine chronometer, which he worked to perfect over the next thirty years, eventually achieving the accuracy required for the prize. Although he was eventually paid over twenty thousand pounds for his thirty-year effort, as Dava Sobel described in her masterly work, *Longitude*, he was never officially acknowledged by Britain's Board of Longitude as the winner of their award.

The fact that clocks could now be synchronized with relatively high accuracy in the far corners of the world created a truly universal terrestrial timeframe. Indeed, today we refer to the local time at the Royal Observatory in Greenwich as Coordinated Universal Time.

Prior to the introduction of standard time at Greenwich, local municipalities set their own time based on the position of the sun at their location. However, once rail travel was introduced, it became possible to travel swiftly enough over long enough distances that clocks on the train had to be reset as the train passed each town.

In a sense, therefore, it was the onset of train travel that led to the standardization of time. In the nineteenth century, train travel began to require coordinated time between villages, and this likely sparked Albert Einstein's interest in the measurement of time. Einstein was a patent clerk in Bern, Switzerland, a country where a train left the station in almost every city every few minutes and where the trains are still famously almost never late. I used to visit the University of

4

Zurich each summer (where Einstein eventually received his PhD), and the old saying that one could set one's clock by the trains was absolutely true—a great advantage in the era before cell phones and Apple watches.

As in almost all of his work, Einstein began by questioning the validity of a key assumption that most people had taken for granted—in this case, that personal time is universal time. He took the sentiment later expressed by George Harrison and the statement I made at the beginning of this chapter and elevated it to a postulate: the only time we can measure is the time we experience right where we are.

This may sound like a tautology, but our experience in the world has conditioned us to assume that the time measured on our watch must also be the time measured in the next room. This is an assumption that needs to be empirically checked. And as Einstein recognized by examining precisely how one might go about doing this, the assumption can be wrong.

While we directly experience time at our own location, all knowledge about the flow of time elsewhere comes to us from information we receive, which takes a finite time to get to us from remote locations. We then make inferences based on these remote observations.

Because the speed of light is so great, when we observe events around us, it seems natural to assume that they are happening contemporaneously, since for all intents and purposes, the time delay between the occurrence of these events and our observations of them is imperceptible. Einstein was driven to question this common-sense assumption because of his realization that the forefront physics of his era contained a paradox.

The theory of electromagnetism had just been fully exposited by the great theoretical physicist James Clerk Maxwell only forty years earlier, based on the groundbreaking work of the equally great experimental physicist Michael Faraday. Maxwell's theory predicted that light is a wave of electromagnetic fields whose speed is determined by two fundamental constants of nature: the strength of electricity and the strength of magnetism. These constants reflect the underlying properties of space itself, and therefore they should be measured to have the same value for all observers.

Einstein recognized that Maxwell's result would imply that all observers would have to measure light as having the same speed relative to them, regardless of their own state of motion, either toward or away from the source of light they were observing. If this wasn't the case, the properties of electric and magnetic forces would be measured differently by different observers, but that would conflict with the universality of Maxwell's theory.

Einstein opted to assume that Maxwell's theory was fundamental and not observer-dependent. But he realized this created a problem, because common sense tells you that if you are moving toward the source of a light beam, the beam should appear to be moving toward you faster than it would be if you were standing still, just as oncoming traffic appears to be approaching you faster than it would if you were standing still at the side of the road.

To his immense credit, Einstein was willing to ask what the consequences would be if common sense was, in this case, wrong. Since speed is determined by distance traveled in a specified time, he discovered it would be possible to reconcile Maxwell's prediction that the speed of light should be observer-independent if instead, measurements of distance and time *were* observer-dependent. If somehow distances and times *both* change in a coordinated way for two different observers in relative motion, the measured speed of light can be the same for both of them, and the universality of Maxwell's theory can be vindicated.

In making this rather dramatic leap, Einstein realized that while this assumption appeared to conflict with experience, because the speed of light is so great compared to the speeds normally experienced by humans, any expected variations in measurements of distance and time for different observers on earth would be imperceptible. Thus, such changes could potentially escape detection.

Once he made his assumption that space and time were relative, it turned out to require nothing more than high school algebra to calculate precisely what the magnitude of the variation in space and time measurements between two observers in relative motion should be.

While the math was ultimately simple, setting the physical problem up in the correct way in order to derive his equations required some imagination. Einstein resorted to what he called *Gedanken experiment*, German for "thought experiments," in order to calculate the differences in measurements of space and time between two observers in relative motion. Living in Switzerland, it was natural that the thought experiments Einstein used to guide his analysis involved trains and, in particular, clocks on trains. He imagined how an observer at the station platform would measure the ticks of a clock seen on a moving train and how an observer on the ground, who synchronized his clock with an observer at the center of a moving train, would measure the time on clocks at the front and back of the train that were synchronized with the clock of the observer on the train. Finally, he considered how an observer on the station platform would measure the length of the train itself.

Any introductory textbook in physics repeats the thought experiment calculations he carried out, but for our purposes it is sufficient to state his results:

1. Clocks observed on the moving train by an observer on the platform will appear to be running slowly.
2. Clocks at the front and back of a moving train that were synchronized with the clock at the center of the train will not appear to be synchronized when viewed by an observer on the station platform. The implication of this is that the time-ordering of events that happen at locations that are spatially removed from each of two different observers can be different for each observer. For such events, one observer's "before" is another observer's "after."

3. The length of a ruler held in the direction of the train's motion by an observer on the train will appear shorter when measured by an observer on the station platform.

In all these cases, when the velocities, v, are small compared to the speed of light, c, the magnitude of the measured differences in observations made by the two observers is on the order of v^2/c^2. This is a very small number, so it was natural that the effect would have remained unnoticed at the time Einstein carried out his analysis.

Now that we can measure such differences, Einstein's predictions have been verified. Time and space are relative, and "now" only has an objective meaning for events that happen where you are located, so "now" is not a universal concept throughout the universe.

While Einstein's predictions are strange, they do not seem immediately paradoxical—that is, until you consider what the observer on the train measures. That observer measures precisely the same effects observed by the observer on the platform, but now applied to clocks and rulers on the platform! *Both* observers measure the other's clocks to be running slowly, the rulers held by the other to be shorter, and so on. The effects are completely reciprocal.

When first hearing this fact, most people infer that the differences in measurements between the two observers are just an illusion and do not reflect any objective reality. How is it possible for my clock to be running slow as measured by your clock, but your clock to be running slow as measured by my clock?

This, however, is only a paradox if you assume that the flow of time is universal and that your measurements have any objective significance beyond your local frame. They don't. The flow of time *is* observer-dependent. To demonstrate that these measurement differences are not merely an illusion but are real, we can resort to a famous example first described by Einstein: the so-called twin paradox.

Consider two twins, one of whom sets off on a spaceship at near the speed of light on a round trip voyage to a star twenty-five light years away. Fifty years later the twin who stayed on earth welcomes his brother home, only to discover that his brother has hardly aged at all, while he is half a century older!

At first, this might not seem paradoxical. After all, as observed on earth, the twin in the spacecraft has a clock that is running slowly, so it may have only recorded the equivalent of a week during the voyage.

The problem arises when one considers the situation from the frame of the brother on the spacecraft. Won't that brother measure the clock of his earth-bound brother to be running slowly as well?

The resolution of the paradox relies on the fact that the situation for the two brothers is not reciprocal because the brother in the spacecraft is not moving at a constant speed during the entire voyage. In order to turn around and come back, the brother on the spaceship must slow down, come to a stop, and reverse. During that period, he experiences a deceleration and then an acceleration, alternately pulling him from his seat and pushing him back into his seat. (Alternatively, the brother can swing around the star and return, but in that case too, he also experiences an acceleration). The brother on the ground, however, experiences no such accelerations.

Clearly then, something strange must happen as a result of acceleration. Indeed, when one works out the mathematics, almost all of the aging for the brother on earth takes place during the short time his astronaut brother is turning around. Before the astronaut turns around, the clock of the brother on earth is running behind the clock on the ship, but after he turns around, when he peers at earth with the powerful telescope on his ship, he sees that the date recorded on the terrestrial clock is now almost fifty years ahead of his clock!

Lest you think this is all hocus pocus, the result of the twin paradox has been tested using a sensitive atomic clock aboard an aircraft orbiting earth. When it returned to base, the clock on the plane was indeed behind that of an equivalent atomic clock on the ground—in this case, only by a few millionths of a second, but enough to confirm the prediction of Einstein.

The strangeness of the flow of time for an accelerating observer probably provoked Einstein's curiosity about accelerating systems. Using a very different type of thought experiment that I shall discuss at length shortly, Einstein convinced himself that whatever happens to someone who is accelerating should happen identically to someone experiencing a gravitational force (as the twin in the spacecraft would feel if he swung around the star). In short, Einstein theorized that all measurements made in the frame of an accelerating observer would be identical to measurements made if the observer was at rest but in a gravitational field.

This was the steppingstone that led Einstein to his general theory of relativity a decade later—his theory of gravity. We do not have to get into all the details of that theory yet. Suffice it to say that this relativity of time argument implies that if acceleration changes the flow of time in objectively measurable ways, then so must gravity.

The first test of this was done by Robert Pound and Glen Rebka Jr. at Harvard, in 1959–1960. These two ingenious experimenters placed a radioactive cobalt source near the roof of the Harvard physics building, which emitted energetic gamma rays that were absorbed by a sample of iron-57 (^{57}Fe). In their experiment, the Fe sample would become excited and subsequently emit gamma rays of a very specific energy, and hence a very specific frequency, corresponding to the energy difference between the ground state of iron nuclei and their first excited state. At the other end of a long tube that ended in the basement of the building seventy-four feet below was a similar Fe sample, along with a gamma ray detector. If the frequency of gamma radiation emitted at the roof was the same as the frequency of the radiation impinging on the Fe source in the basement, that source would absorb the gamma rays to energize nuclei into their first excited state with high efficiency.

One can think of the frequency of light as a precise clock, as the frequency in cycles per second is like the ticking of a clock. In the case of the gamma rays emitted when Fe nuclei relax from their first excited state, there are more than a million trillion ticks each second.

If clocks in the basement ticked at a different rate than clocks on the roof, then the frequency of radiation emitted or absorbed by the sources on the roof and basement would be different. Pound and Rebka were able to confirm this very small effect by moving the Fe source on the roof up and down vertically. Due to the famous Doppler effect, whereby the frequency of radiation from a source moving toward you is higher than it would be if it were at rest relative to you, while the frequency due to a source moving away from you would be lower, Pound and Rebka were thus able to slightly vary the frequency of light emitted by moving the source on the roof with respect to sources at rest in the laboratory.

Sure enough, when they varied these frequencies, they found that the Fe source in the basement preferentially absorbed light emitted by the moving Fe source on the roof when it had a slightly smaller frequency than the frequency that would have been emitted if it had been at rest with respect to the absorber in the basement. Similarly, by putting the moving source in the basement, they found that the Fe on the roof preferentially absorbed light emitted with a slightly higher frequency than would be emitted by an Fe source at rest in the basement. This meant the Fe "clock" in the basement was ticking more slowly than the Fe clock on the roof, just as Einstein had predicted!

Since the net effect predicted by Einstein was only slightly more than one part in a million billion, this experiment represented a triumph of experimental ingenuity in 1960. Nowadays we have very sophisticated atomic clocks with accuracies far exceeding those needed to test this prediction.

Since the gravitational field of earth is not large, we can understand this general relativistic effect in somewhat simpler terms: in climbing up against the gravitational pull of earth, the light loses energy. Lower frequency light has less energy and thus the light moves to longer wavelengths and lower frequencies. This effect is called a gravitational redshift, because red light is the longer wavelength part of the visible spectrum.

While this effect is very small, as I have mentioned, modern technology allows it to be directly measured—so much so, in fact, that this gravitational redshifting actually plays a large role in governing our everyday lives. If you have ever used the GPS on your phone to guide you as you drive or walk, you have relied on the fact that we know about gravitational redshift and can adjust our atomic clocks accordingly.

GPS satellites work by using a triangulation procedure that we can schematically simplify as follows. They carry carefully calibrated atomic clocks and can send out time signals that are picked up by your phone, which can in turn record the time the signal is received. This tells your phone how far away it is from the satellite since the signal travels at the speed of light. If your phone can do the same thing with three or more satellites, then it is possible to triangulate its position precisely in three dimensions.

All of this of course depends on the careful calibration of the different clocks. However, the various satellites are moving relative to you at different and relatively high speeds, and their altitude is about twelve thousand miles above earth's surface. Together these factors mean that the clocks on the satellites are observed to tick at slightly different rates than an identical clock on earth would tick.

In this case, due to the effects of special relativity, the speed of the satellites slows their tick rate by about seven microseconds each day, while due to general relativity, their higher altitude causes their tick rate to speed up by about forty-five microseconds each day. This may not seem like a lot, but if we didn't take that into account, our location accuracy would be off by almost a kilometer after an hour of use.

If the rate at which time flows depends not just on the motion of an object, but the environment in which it finds itself, then it is natural to consider more exotic environments and what they may do to time.

But how exotic can the universe get? Are there places where time itself doesn't exist? That, we do not yet know....

* * *

The first occasion where we approach this limit of the known occurs when gravity gets so strong that our ability to observationally probe it is limited, and our fundamental understanding of gravity itself may break down.

As I have described, my clock on earth will be measured to be ticking more slowly, by a very small amount, than a clock in a satellite at a higher altitude. But earth has a relatively weak gravitational field. What about more massive objects, where gravity at the surface is much stronger?

The first person to seriously consider the extremes of gravity was an unheralded British clergyman and scientist, John Michell, who was born three years before the death of Isaac Newton. He taught geometry, Greek, Hebrew, philosophy, and geology at Cambridge, where Newton had unveiled his famous universal law of gravity only seventy-five years earlier. One historian of science, Sir Edmund Whittaker, has argued that Michell was the only natural philosopher of merit at Cambridge in the century following Newton, but alas, history was not kind to Michell, and as Whittaker put it, his name "perished entirely from Cambridge tradition."

Be that as it may, in 1783, almost one hundred years after Newton's *Principia* was published, Michell first proposed what he called "dark stars." Newton had assumed that light was made of particles, and Michell took that hypothesis and ran with it. He reasoned that like cannonballs and apples, particles of light would be attracted by the gravitational pull of a planet or star and could be pulled back toward the planet or star if their speed was not high enough to escape.

The escape velocity of objects launched at earth's surface was known even then to be about eleven kilometers per second (or about seven miles per second). But what about heavier objects, like the sun?

13

Newton himself had estimated the ratio of the mass of the sun to that of the earth to be about two hundred thousand to one, about a factor of two smaller than the actual value. This distance to the sun was measured by using data from the transits of Venus in 1761 and 1769. Then, from the angular size of the sun in the sky, it was possible to determine its radius. Once one has the radius and the ratio of the mass of the sun to that on earth, one can use the measured acceleration of objects at the earth's surface to obtain the escape velocity from the surface of the sun. This is about 618 kilometers per second, almost sixty times larger than the escape velocity from the earth, and about one five-hundredth the speed of light.

Michell then wondered about objects heavier than the sun. He considered a star like the sun, with the same composition and thus the same density as the sun, but scaled up in size. The escape velocity in this case scales directly as its radial size. Thus, Michell calculated that a star five hundred times the size of our sun would have an escape velocity equal to the speed of light.

He argued the universe could be full of many of these dark stars. As he so presciently put it in 1783:

> "If there should really exist in nature any bodies, whose density is not less than that of the Sun, and whose diameters are more than 500 times the diameter of the Sun, since their light could not arrive at us; or if there should exist any other bodies of a somewhat smaller size, which are not naturally luminous; of the existence of bodies under either of these circumstances, we could have no information from sight; yet, if any other luminous bodies should happen to revolve about them we might still perhaps from the motions of these revolving bodies infer the existence of the central ones with some degree of probability."

We now know that Newton's law of gravity is not the correct theory to use when escape velocities approach the speed of light. Instead, we must resort to general relativity, which allows for the curvature of space as well as time dilation. Nevertheless, general relativity gives precisely the same answer for the radius where the escape velocity is the speed of light, which is called the event horizon. So, Michell was on the right track, and today we call such objects not dark stars, but black holes.

Michell's analysis was, in retrospect, even more remarkable. He suggested we might discover the existence of dark stars by observing the motion of bodies orbiting around them. That is precisely the tool that astronomers used to confirm the existence of a black hole at the center of our galaxy; moreover, the observation was sufficiently important that the 2020 Nobel Prize was given for it. Not bad for a now forgotten Cambridge professor, who also developed the first experimental device to measure the strength of gravity, although it was left to others to use it after he died.

Unfortunately, Michell's precocious prediction disappeared into the dustbin of history, and the question only resurfaced as the possible existence of black holes in general relativity began to be debated. Einstein himself dismissed the possibility because of the same worry about the implications for understanding the laws of physics that we shall discuss here. It took almost fifty years before physicists had resigned themselves to even the theoretical *possibility* of real black holes, and another twenty-five before definitive evidence of the existence of astrophysical black hole-like objects was obtained.

What makes general relativity black holes much more interesting and more mysterious than Michell's dark stars is that not only does the escape velocity of a black hole increase as objects become more massive, but space and time are also both noticeably altered in the process. I shall have more to say about space in the next chapter. Here I will concentrate on time.

As we have seen, clocks located deeper in a gravitational potential well, like that of the earth, tick more slowly relative to clocks located further outside of such systems. As the gravitational potential well gets deeper (and correspondingly, the escape velocity increases), this effect becomes more pronounced. Eventually, once the event horizon is formed, time will appear to stop completely.

From an operational perspective, consider a person falling in toward the event horizon of a large black hole and emitting an SOS signal by waving a flashlight at some regular rate. As they approach the event horizon, the time between the flashes you might see will get longer. You can think of this as the ticking of a clock that gets slower and slower. But more than that, the wavelength of the radiation in each flash will also get longer as the waves get stretched out, rising up out of the potential well. The light in each flash will go from blue, say, to yellow, to orange, to red—and after that to infrared, to microwaves, and to radio waves.

This combination makes things particularly interesting. The first fact leads to an apparent paradox. You could actually never see someone fall into a black hole because they would appear to fall more and more slowly due to the slowing of clocks. In their own timeframe they would cross the event horizon without noticing anything strange. But for an outside observer, they would appear to freeze just outside the event horizon. For that reason, in Russian, black holes were known as "frozen stars." (It's a less catchy name of course, which is perhaps why there are no Russian science fiction films featuring them.) But an outside observer won't actually see this freezing, because the light from the infalling victim's flashlight will be redshifted to ever longer wavelengths until it is literally undetectable. The person will thus disappear from view before they actually cross the event horizon.

This becomes a bit more problematic if one accepts the calculation, first done by Stephen Hawking, that when one incorporates the effects of quantum mechanics (about which I shall speak later) into black hole physics, then black holes would actually radiate away their energy as if they are objects existing at some non-zero temperature. As they do this, they continue to get hotter, radiating faster, until in principle they radiate away completely.

The time it would actually take for a macroscopic, say solar-sized, black hole to radiate would be immense—far longer than the current age of the universe. However, this long but still finite time does present a problem. From the point of view of a distant outside observer, it would take an infinite amount of time to observe the formation of a black hole, since as the infalling material got closer and closer to forming the black hole, the slower and slower this material would appear to be falling until it would appear to hover close to the emerging event horizon. But, from the point of view of this same distant (long-lived) outside observer, the black hole would radiate away in a finite time. So in this frame, the black hole would disappear before it was observed to fully form.

This is problematic, but it is not impossible. Indeed, it suggests that somehow a record of everything that falls into a black hole throughout its entire formation history and existence may be stored somewhere near the event horizon. We will come back to that possibility in the next chapter, where we will focus on space and black holes.

For the purposes of considering questions associated with time, what this strange behavior makes clear is that the behavior of time near the event horizon needs to be explored further. It also reveals that time, as measured by an observer outside a black hole, must be very different than time as measured by an observer who has actually crossed the event horizon.

It is natural to suspect, since time appears to slow to zero at the event horizon, that time actually reverses direction inside of a black hole. But this is not the case. What happens is stranger still.

In special relativity and general relativity, the distinction between space as separate from time disappears. They are combined together into a four-dimensional "spacetime." Choosing what corresponds to the time direction or space direction will in general be observer-dependent. This is a generalization of the more familiar fact that in three-dimensional space, defining what direction is "up" is observer-dependent. Observers in Australia who point up at the sky will be pointing in the opposite direction from observers in Europe. Similarly, in spacetime it is possible that one person's perceived spatial direction can be another person's time direction.

This is essentially what happens as one traverses the event horizon of a black hole. To understand how this arises, consider the following. We can distinguish the past from the future using a formalism elaborated by Roger Penrose. Our past "light cone" contains all possible light signals that could have been received at any time from the beginning of time until the moment we call the present. The future light cone includes all the places that I can communicate with in the future by sending out a light ray at the present time.

Now, what happens as I get closer to the event horizon of a black hole? Due to the curvature of space around the black hole, rays of light that I send out in any direction begin to curve in the direction of the event horizon. As I get closer and closer to the event horizon, more and more of the light rays bend until they are pointing directly toward the spherical event horizon surface. At the event horizon, all light rays I send out will bend inward, toward the inside of the black hole. As I cross the event horizon, my entire future light cone points radially inwards. I can't go up. I can only go down. My future position has only one direction—downward—with a decreasing radius corresponding to a more distant future (although as we will see, that future will not be too long). Before I entered the event horizon, my movement in time was always only forward. Now my future is only downward. Space (aka radius) has become timelike.

What about space? As I fall inward, I will detect rays of light that appear to arise from below that may have entered the horizon at the earliest moments of the formation of the black hole. For a hypothetical black hole, these may be arbitrarily far back in time. From "above," I will encounter light rays that entered the event horizon after I did, and because of blue shifting, these will have entered the event horizon arbitrarily far in the future, as time ticks in the outside universe.

TIME

Looking in one direction, I see the past. Looking upward, I see the future pass before my eyes. Past and future will be distinguished by their direction, and both directions appear to be accessible. Time has become spacelike. I can distinguish and access both past and future as I move about before my own end, which is written in stone as I inexorably fall, decreasing in radius no matter what I do. Indeed, in one of the bitter pills associated with black hole physics, the harder I try to move up, the faster I end up moving down.

From this perspective, this temporal "space" inside a black hole can be arbitrarily large, even though as seen from the outside, the event horizon appears to encompass only a finite, and potentially small, volume. It could be large enough to contain a whole, in principle infinite, and perhaps even expanding universe. The inside of a black hole is in this sense reminiscent of the wardrobe in C.S. Lewis' *The Lion, the Witch, and the Wardrobe*. In this book, a wardrobe of finite size as seen from the outside allows access to a whole new world contained inside.

Time, on the other hand, is finite. The inexorable march down in radius ends at a place where space and time appear to lose all meaning. This "singularity" is a point where the physics we understand breaks down. It is commonplace to think of it as a point of infinite density, but it rather represents a finite point in time potentially stretching out infinitely in space.

For the infalling observer, the singularity represents the end of time. Time stops there, gravitational forces become extreme, and there is no way to avoid this end once one enters the event horizon. For someone falling into a black hole formed from an object with the mass of our sun, the proper time between entering and encountering the singularity would be less than the blink of an eye. For supermassive black holes of billions of solar masses, as appear to exist at the center of many galaxies, one might have a minute or so to contemplate the end.

But what kind of end is it? We don't know. Once notions of space and time break down, our ability to picture phenomena, to describe events, and to make predictions disappears.

We don't know if there *is* a singularity, or whether a new understanding of the fundamental physics of gravity will change the behavior of space and time at small scales. Most physicists are betting on, or perhaps hoping for, such a change, but the universe doesn't exist to make physicists happy.

We don't know what happens to time, or space, if there is a singularity.

We don't even really know what the end of time even means.

There are a host of speculations about this ultimate state of black hole collapse and its physical manifestations, with perhaps the most optimistic being that one can pass through the singularity into another universe of space and time. But until we have a theory of gravity that remains valid at the extremes of curvature and the infinitesimal scales of space and time present near the singularity of a black hole, it all remains speculation.

* * *

We all eventually have to come to grips with the end of our own time—our death. The prospect is sufficiently terrifying that many of the world's religions provide reassurance through claims that resemble some of the speculations about the time singularity of a black hole. They argue that death transports humans to a different realm, an afterlife, where time as we experience it does not exist. Many also explicitly forecast an "end of times," where the existence of our worldly stage will disappear. In this sense, the possibility of a finite end of time in a black hole may not seem unfathomable.

But a beginning of time is far more challenging. It begs for metaphysics. After all, if there was no "before," then what does one make of cause and effect—the centerpiece of our experience of the world? If time itself begins, then nothing precedes the emergence of existence and the dynamics of our world, and there thus appears to be no proximate cause of our very existence, or at least no natural cause. Not surprisingly, this has led some to fall back on a last refuge that allows them to avoid the truly difficult questions, namely God. But for the rest of us, dealing with a possible true beginning of time forces us to confront a series of challenges for physics.

This is the dilemma that has recently faced the science of cosmology, a discipline where much of my own professional career has been centered. I say recently because even a century ago, there was no apparent problem. The perception among astronomers was that the universe was essentially static and eternal, with no beginning or end. There was no evidence of any large-scale evolution of the then-visible universe, so this was not an unreasonable assumption.

In 1929, however, Edwin Hubble, combining his own data taken at the Mount Wilson telescope with data taken by others, provided evidence that light from distant galaxies was being shifted to progressively longer wavelengths with increasing distance from us. This cosmic "redshift" (which, as I described, is named because red light represents the longer wavelength end of the visible spectrum) could be most simply interpreted as being due to the Doppler effect for moving objects, whereby the wavelength of light from receding objects is shifted to a longer wavelength. Taken at face value, it implied that the greater the distance between galaxies, the greater their relative speed of separation.

Hubble described his result as reflecting "apparent velocities" of recession of galaxies that were proportionate to the distance separating us from the galaxies, but he was agnostic about whether these in fact represented real velocities, or some other effect.

In hindsight, the simplest interpretation of Hubble's results is that the universe is uniformly expanding in all directions. Hubble, however, never really accepted this result, in spite of the fact that a Belgian priest and physicist, Georges Lemaître, had actually predicted this very phenomenon two years before Hubble announced his result.

At the time, Lemaître was a little-known part-time lecturer in Belgium, but he demonstrated that general relativity allowed, as a cosmological solution, a uniformly expanding universe. His groundbreaking paper appeared in a little-read journal, so it was not widely noticed until the British astrophysicist Arthur Eddington translated it into English in 1931.

The notion that the universe could be expanding was at the time so heretical that even Albert Einstein, whose equations predicted such a possibility, refused to accept it, uttering his famous comment, "Your calculations are correct, but your physics is atrocious."

In 1931, thanks to Eddington, Lemaître's work had become well-known, and he had also responded to Einstein's concerns. At that point, he made what seemed like a bolder proposal, which again, with hindsight seems inevitable: if the universe is expanding, then in the past it was smaller. Extrapolating this expansion backward, at some finite time in the past, the entire universe would have comprised a single infinitesimal point, which Lemaître dubbed the "Primeval Atom" and the skeptical scientist Fred Hoyle facetiously referred to as the "Big Bang" in 1949.

While Eddington found Lemaître's expanding cosmology brilliantly consistent with Hubble's observations, the inevitable consequence of extrapolating it backward was far less appealing to him. Einstein objected to the emergence of the universe from such an infinitely dense singular point on physical grounds for the same reason he had opposed black holes.

But Eddington and Einstein aside, once again we need to remember that the universe isn't governed by what scientists may find appealing or unappealing. A classical extrapolation backward of the currently observed Big Bang expansion implies that approximately 13.8 billion years ago, the universe emerged from a singular point. Just like the singularity at the last stage of black hole evaporation, time and space literally break down at this point.

Roger Penrose demonstrated in 1965 that within the context of general relativity, the final stage of black hole collapse must produce a singularity, and he won the Nobel Prize for his work. Stephen Hawking later expanded on Penrose's demonstration to show that under certain general conditions for the properties of energy such as exist in a universe dominated by matter or radiation, following the equations of general relativity backward inevitably leads to a singularity in the finite past at which time and space cannot be defined. In short, there appears to be a time in which there was no "before," at least in any way that we currently understand the word.

The Big Bang has become so ingrained in popular culture that is hard to fathom the psychological shift that accompanied the realization that the universe might not be static and eternal. An eternal universe begs all the hard questions. No worries about the need for creation or the future. No concerns about why life on earth has evolved *now*. If there is a beginning, however, everything changes.

Perhaps for this reason, there has been pushback from many fronts, from the time Lemaître first made his proposal up until the present time. The first criticism came from British scientist and science fiction writer Fred Hoyle, who derided the idea by giving it a name that he thought was facetious. Unfortunately, *the Big Bang* was such a good name that it stuck. In any case, he, along with his colleagues, worked until the end of his life to try and develop a counter theory, which they called the steady-state cosmology in which the universe is eternal and unvarying on its largest scales.

While Hoyle could debate the significance of the observed Hubble expansion, the 1965 discovery of the Cosmic Microwave Background (CMB), which is due to the radiative afterglow of the Big Bang, dealt a fatal blow to the steady-state theory. The discovery provided a firm empirical basis for not only the reality of the Big Bang expansion itself, but our ability to extrapolate the observed expansion backward in time billions of years, to just three hundred eighty thousand years after the Big Bang, the time when the currently observed background began to evolve on its own.

A year later, James Peebles, who was one member of a group at Princeton University that had set out to discover the CMB, only to be scooped by two researchers at nearby Bell Laboratories who discovered it by accident, made a robust prediction that demonstrated that the Big Bang expansion could be reliably extrapolated all the way back to mere seconds after it occurred.

In the 1940s, Ralph Alpher and George Gamow took the Big Bang idea seriously enough to recognize that extrapolating it backwards could imply that the early universe was not only denser, but hotter. In the earliest seconds, the temperatures could have exceeded ten billion degrees. At these temperatures and densities, nuclear reactions would have occurred, and they recognized that these could have allowed an early dense plasma of protons, neutrons, electrons, neutrinos, and radiation to evolve, producing heavier elements like helium and lithium.

After the discovery of the CMB confirmed the hot Big Bang picture, Peebles was able to demonstrate, based on both the measured temperature of the CMB and nuclear reaction rates as measured in the laboratory, that generically, about 25 percent of the protons in the universe, by weight, would have reacted to form the nuclei of helium in the first few hundred seconds of the Big Bang. At that time (and indeed since that time) no models of the evolution of stars in the universe have been able to explain how to convert more than about 2 percent of primeval protons into helium in the nuclear reactions in the cores of stars. The measured abundance of helium in the oldest stars and interstellar gas, however, indeed turns out to be about 25 percent. A stunning prediction with no other possible explanation than the Big Bang.

Since that first calculation, measurements of the cosmic abundance of light elements, including deuterium and lithium, whose predicted values range from one part in a hundred thousand for deuterium to one part in a trillion for lithium, also agree with Big Bang predictions. Moreover, the precise predictions depend on the density of protons in the universe, and this density can be independently measured from the detailed properties of the CMB; and you guessed it, the predicted proton density from Big Bang nucleosynthesis agrees bang on with the CMB measured value.

In short, not only did the Big Bang really happen, but the agreement between predictions and observations implies that we can reliably extrapolate the current expansion back to a second or so after the Big Bang itself.

A second doesn't seem like a lot, but that is because we normally live our lives in minutes and hours. Of course, remember that admonition by Mr. Relativity, Albert Einstein: if you are sitting on a hot stove, a second can seem like an hour...

Less facetiously, while we tend to experience time linearly, there is a sense in which one second after the Big Bang is infinitely far from t=0. This is because the rates at which physical processes occur tend to depend on the temperature of the universe, so that, at high temperature and density, reaction rates are exponentially faster than at low temperatures.

The temperature of the universe varies inversely as a power of time, so that as time goes to zero, the temperature of the universe goes to infinity. Thinking in terms of powers of ten, there are an infinite number of powers of ten between ten billion degrees and infinity. Or, thinking in terms of powers of ten in time, there are an infinite number of (negative) powers of ten between one and zero.

As an interesting aside, here is a little-known fact: Since reaction rates increase exponentially with temperature, and temperature skyrockets as one approaches t=0, it is possible to estimate that more reactions occurred between particles in the first second of the history of the universe than will likely occur in the entire future history of the universe, even if that history is eternal! In this sense, we were born after almost all the good stuff happened.

As fascinating as this fact may be, I want to make one thing clear at this point, because I get a lot of mail asking about this. Would clocks have ticked at the same rate in the earliest moments of the Big Bang, or would they too have sped up, or maybe even slowed down? The answer is: they would have ticked at the same rate as our clocks on earth do now, more or less. The first second really was a second long, or at least almost all of the first second was almost all of a second long. Once we get back to very close to t=0, all bets are off because our current understanding of the laws of physics at that time breaks down. (More on that in a bit, because it gives a possible way out of Hawking's proof of a finite past.) But after that presumed instant, even though the universe was hot and dense, for all "co-moving observers"—that is, observers who are locally at rest and being carried along with the expanding universe—the rate at which clocks tick doesn't change. So, the cosmic time we label as one second in the Big Bang really is approximately one second. I put approximately here because of our uncertainty about precisely what happens at or very near t=0.

I expect some science fiction writer must have imagined a hypothetical observer at those early times whose metabolic rates scaled with the temperature of the universe. If we think of the number of events experienced as the marker of a long life, such an observer might feel like they had lived for almost an eternity if they survived from the earliest moments of the Big Bang through to the ripe old age of one second.

Returning to the thrust of our discussion, however, the fact that there are an infinite number of negative powers of ten between zero and one second illustrates the real challenge of ever getting direct empirical probes of the moment of creation itself. Also, as reaction rates generally scale with temperature and thus as inverse powers of time, one can imagine a host of possible significant events that could have unfolded between t=0 and t=1 second, any one of which could have erased interesting remnants from earlier times that one might hope to somehow still detect today.

In fact, we already know that a host of interesting phenomena occurred before the universe was a second old. When the universe was about a millionth of a second old, the quarks that are the fundamental particles making up protons and neutrons would have first obtained their masses. At earlier times they would have behaved like essentially massless particles. Coincident with this, around this time quarks would have first become confined into the particles we are familiar with today: protons and neutrons.

A million times earlier, two of the four forces of nature, the weak and electromagnetic interactions, would first have begun to diverge in character. Before that, they would have been essentially indistinguishable. One can imagine other possible cosmic milestones that occurred even earlier, but our ability to directly probe such physics at earlier times is so far limited by our ability to build large accelerators. The largest accelerator in the world, the Large Hadron Collider in Geneva, Switzerland, currently probes the electroweak scale, and that is as far as we can directly go with experiment today.

Having said this, the possibility I mentioned earlier that some phenomenon that took place at early times might have wiped out all evidence about what preceded took on a new dimension as the result of a revolutionary idea that was proposed in 1980 and has since become the centerpiece of modern cosmology. I refer to cosmic inflation.

The physicist Alan Guth realized, when thinking about the particle physics of the early universe, that a phenomenon analogous to the moment when the weak and electromagnetic forces first began to diverge in character in the early universe could have occurred even earlier if the three non-gravitational forces in nature were once unified in what has become known as Grand Unified Theory. Generally, during such a transition, which physicists call a "phase transition," a new kind of behavior would have governed the expansion of the universe.

For example, on an urban road, with lots of traffic, water will not freeze on the pavement even if the temperature is below 32 degrees Fahrenheit because it is constantly agitated by the passing traffic. Once the traffic slows, however, the water can suddenly freeze, forming black ice. The water freezing below its natural melting point releases energy, which keeps the newly formed ice from cooling further for some time.

A similar phenomenon can occur in the universe. As it expands, a cosmic field, similar in nature to the Higgs field that currently permeates all of space and endows elementary particles with mass, can get "stuck" in what we call a false vacuum—a state that is not its state of minimum energy. Eventually it can settle down to its true minimum, releasing the energy it stored before this transition. In general relativity, this type of energy stored in space turns out to be gravitationally repulsive, not gravitationally attractive like all other kinds of energy.

Guth realized that when this happens, it could cause a sudden expansion of the universe, which could cause space to expand by many orders of magnitude in an instant shortly following the Big Bang, a phenomenon he dubbed inflation. He recognized that this process of inflation could solve several longstanding problems in cosmology. To date, it is not only a natural explanation of why the universe appears the way it does today, but it is the *only* explanation that relies on well-defined physical ideas.

This is not the place to discuss the fine details of inflationary theory. I have done that in my book *A Universe from Nothing*. Rather, of interest here is what inflation would imply for our ideas about time—in particular, the beginning of time.

The first and somewhat depressing implication of the theory is that following inflation—after the phase transition completed and the energy stored in space was released to produce the initial conditions of the subsequent hot Big Bang expansion—essentially all traces of the pre-inflationary state of the universe were erased. This means that if we were hoping to observe some remnant signal today that could provide empirical information about the moment when time may have begun, the chances of such a possibility are now exponentially more remote than they were before, and they were remote even before inflation was posited.

The next implication is more interesting. It turns out that inflation doesn't end easily, at least not everywhere. Locally, the background field can relax from its metastable state, and the region in which it relaxes will stop expanding exponentially quickly and can heat up. This creates the initial conditions for what, in that region, will be observed to be a hot Big Bang expansion. But because space expands exponentially in the false vacuum regions between the regions where the transition has occurred, those latter regions make up a smaller and smaller fraction of space as a whole. The technical term for this is that the phase transition never "percolates"—it never grows to encompass all of space.

As the physicist Andrei Linde, one of the other founders of the modern theory of inflation, realized, this generically makes inflation "eternal." The space in which the metastable field is trapped in a false vacuum state just keeps growing forever. The small islands in which a transition has taken place decouple from the background exponential expansion and undergo their own separate evolution. Each has what appears to be a "moment of creation"—the moment when the transition occurs that the background field energy gets released as heat energy, initiating a local hot Big Bang expansion—but that moment can be arbitrarily far from the instant when space and time themselves may have first arisen globally, if such an instant exists.

Another implication of this phenomenon is the possible existence of what has become known as a multiverse. Each region of true vacuum that nucleates out of the background inflationary expansion acts as a separate universe that is causally disconnected from the rest of the expanding multiverse. Moreover, it is possible that when a phase transition completes in different regions, the ground states in different regions can be slightly different (as, for example when ice crystals form on a windowpane in different directions in different places). If this happens to the configuration of the background field in different "universes," these regions can effectively manifest what will appear to be different forces, particles, and laws of physics. Physics as we know it, then, could be local to our universe and not a global phenomenon.

To return to the issue of time, Woody Allen once said, "Eternity is a long time, especially near the end." But actually, the same is true for the beginning, if it is eternally far away as well. A multiverse that is eternal in the past is very different than one that originated simply a very, very long time before our local Big Bang post-inflation epoch may have begun. It is *infinitely* different, in fact.

Could the multiverse be past eternal, even if our local Big Bang expansion isn't? If we base our conclusions on known laws of physics, including general relativity, then the answer is no. Guth and collaborators demonstrated, barring possible novelties associated with a quantum theory of gravity, that Hawking's arguments about our Big Bang apply to the origin of the multiverse as well. A singularity would appear to inevitably loom in the finite, although possibly incredibly distant, past.

Once one allows for possible quantum effects associated with space and time themselves, however, all bets seem to be off when it comes to the question of an "instant" when time may have begun.

Hawking and his collaborator Jim Hartle proposed early on what they called a "no-boundary boundary" condition of the universe. In this picture, one cannot temporally trace back to a beginning. Space could have emerged without the existence of time at all. Time would have emerged out of what was purely space.

I am tempted to say time would have emerged "after the beginning," but of course that is incorrect if there is no time at all. That is part of the problem of dispensing with time. All our intuitive notions about phenomena no longer apply.

Another possibility, and one I explored myself early on my postdoctoral career only to discover that my nearby colleague and friend, Alex Vilenkin, had independently beat me to the publishing punch, is that inflationary spacetime directly emerged through a quantum process called "tunneling" from "nothing"—that is, no space and no time. A process in quantum field theory called an instanton (named because it can be thought of as a process that occurs at an instant of time) can, in the context of general relativity, describe the quantum sudden emergence of an exponentially expanding space of non-zero size.

Both Hawking/Hartle's and Vilenkin's pictures suggest that quantum gravitational dynamics could both dispense with a troublesome initial spacetime singularity and result in a universe like ours where no such universe existed before—motivating some of the arguments I describe in my book, *A Universe from Nothing*. While the tunneling picture has a more clearly defined "instant" of creation, both proposals also share the property that time as we measure it emerged along with our measurable universe, so that question of what existed "before" our universe may simply not be a valid question.

Is it impossible, then, that eternity is a one-way street, going only into the future and not backward into an eternal past? As you might expect, the answer is no. Once again, the uncertainties of quantum gravity allow a number of possibilities.

Roger Penrose and several other physicists who have made similar independent claims, have proposed—unconvincingly, in my opinion—that our currently expanding universe is simply the latest part of an infinite cycle of expansions followed by contractions. This picture is intuitively attractive both because it gets rid of a beginning and because it treats time symmetrically in the past and future. But intuitive attractiveness is no guarantee of scientific correctness, and so far, the notion of cyclic universes seems convincing to only a handful of physicists. Ultimately, hand-waving appears required to get past the singularity that otherwise seems to separate past contraction from future expansion, as well as some assumptions about the physics of the future that are currently otherwise unmotivated.

Another possibility, which I first learned from Alan Guth recently, is the suggestion that near a quantum singularity in spacetime, time itself may emerge in two directions, so that no clear beginning can be discerned and an infinite regression into the past is possible. Once again, this has some appeal, but is, as far as I can tell, speculation, without any solid basis in theory that I know of.

The result of all this is that quantum gravity promises almost anything but currently gives us very little. It allows us, however, to effectively know what we don't know, by hiding vital details about time as effectively as a black hole hides everything inside its event horizon.

* * *

Lately, especially during Covid, and compounded by my recent move to a rural paradise, my favorite part of traveling has been returning home. That we can return to a location from whence we came is something we all take for granted. Three-dimensional space is fully traversable, forward and backward in every direction.

Not so with time, however. We seem to be inexorably forced, tick by tick, into the future. The mistakes along with the exaltations of the past can be corrected or relived only in our memories.

Seen in the context of general relativity, this dichotomy between space and time seems particularly strange. After all, space and time are entwined together, and as we have seen, one person's time can be another person's space, depending on their frame of reference.

So, what gives? Is a round trip in time possible?

Most of us have probably thought about the possibilities of traveling in time, and some of the best science fiction stories ever written, including H.G. Wells's *The Time Machine*, have dealt with the inevitable paradoxes of visiting another time and changing the past or the future.

These paradoxes demonstrate one of the reasons why time travel is, at the very least, problematic. If I go back in time and kill my grandmother before she gives birth to my mother, then when I return to the present, my mother cannot exist, but then I cannot exist. But then how did I go back in time in the first place?

These kinds of paradoxes often form the grist for science fiction, as in my favorite Star Trek episodes, or the TARDIS from Doctor Who, but there is another problem for a time machine, if one had one, that is less often mentioned. A time machine would simultaneously have to be a space travel device.

Earth is moving in its orbit around the sun at about thirty kilometers per second. If I go back in time just one minute, in my current location, then the earth will be located about eighteen hundred kilometers back in its orbit, about half the distance across North America. An hour back and earth would have been one hundred and eight thousand kilometers back, about one quarter of the distance between the earth and the moon. That means I would emerge from my time machine in empty space—a rude awakening right before a ruder ending.

For these reasons and others that I will elaborate on momentarily, many physicists have assumed traveling backwards in time is impossible. Stephen Hawking, who wrote the foreword for my book, *The Physics of Star Trek*, once said that time travel was impossible, because if it were possible, we would already be inundated by tourists from the future! I responded to him, however, by saying that it was possible they all went back to the wild 1960s and no one noticed.

Nature is the way it is, however, whether we like it or not, and whether time travel is possible will not be determined by whether it causes paradoxes that bother us. And in fact, it is well known that the equations of general relativity do allow for universes where time travel is possible, with "closed time-like curves"—round trips in time. The question of course is whether we live in, or can live in, such a universe.

The equations of general relativity can be put in a particular form that is illuminating. The properties of the geometry of spacetime appear on the left side of the equations, and the properties of matter and energy then appear on the right-hand side. Once expressed in this manner, it is clear that as long as one can write down a geometry equation that contains closed time-like curves (round trips in time) there must be a mathematical configuration of matter and energy that would result in such an outcome. The question, however, is whether such a mathematical configuration of mass and energy is in fact physically achievable. As you might expect, the answer is: we don't know.

As I described in *The Physics of Star Trek*, one such configuration is particularly easy to picture physically: a stable wormhole—basically a tunnel-like shortcut through space that connects otherwise distant points—can always be turned into a time machine. The reason is that if one "mouth" of the wormhole is at rest in the background space, while the other mouth is moving through the background space, then clocks located at either end of the wormhole will be ticking at different rates. One can then rig up a scenario in which one goes through the wormhole, coming out the end with the moving clock, which is ticking more slowly than the clock at the other end of the wormhole, then traveling back through the background space to one's original point of departure, and arriving before one left!

The problem, as Kip Thorne and his collaborators first showed way back in 1988, is that wormholes cannot be stable if normal matter and energy are all we have available to create one. Each mouth of the wormhole will collapse to form a black hole out of which nothing can escape in a time shorter than it would take to traverse the wormhole.

The only way to stabilize a wormhole is to fill it up with a new, exotic kind of energy—basically "negative energy" material—and there are persuasive arguments that one cannot create, even in principle, such energy in the laboratory. Persuasive, but alas not ironclad. Once again, knowing how to untangle the precise relativistic quantum properties of curved spacetimes is required to resolve the issue, and we don't yet have the technology to do so.

The result is that for those of you who yearn for the good old days, there may still be hope....

* * *

Finally, is there actually an end-of-days for our universe? The Bible may favor such a future, but the data suggests otherwise. The observed expansion of the universe appears to be accelerating due to the presence of a non-zero energy permeating all of space.

The acceleration suggests a dismal but eternal future. Eternal, because if this energy of empty space remains constant, then the currently observed expansion will never cease. Dismal, because in this case all the galaxies we now can observe will eventually be receding from us at velocities that exceed the speed of light. (That is allowed in general relativity because in their local frame the galaxies are at rest, and it is the space between us and distant galaxies that is expanding. Special relativity just tells us that objects cannot travel *through* space faster than the speed of light. But the expansion of space itself is not so constrained in the extension of special relativity necessary to deal with such expansion, namely, general relativity.)

In this case, the rest of the universe will essentially disappear on a timescale of a few trillion years, and all that will be left will be our own galaxy (which by then will have collided with several other galaxies and will have become transformed to a roughly elliptical shape, no longer the lovely spiral shape it possesses now). We can consider ourselves lucky to live at a time when we can see so many other galaxies, as astronomers on planets a few trillion years from now will observe only one. Eventually the stars in our galaxy will burn out, and the black hole at the center of our galaxy may grow to devour all the rest of the mass in our galaxy. But if Hawking is correct the black hole will eventually radiate away and disappear much, much later, and all that will be left in the space we now occupy will be a cold, dark, apparently empty universe.

You may have noticed several "ifs" in the last few paragraphs. This dismal future is the future as it "might be," in the spirit of Charles Dickens, because we don't know if the presently observed energy of nothingness—of space itself—is really a fundamental property, or just a transitory one, like the energy that drove a presumed inflationary phase in the early history of our universe.

If that energy disappears, then the future expansion of the universe could be dramatically different, depending both on the unknown geometry of the universe on a scale larger than we can currently measure and the unknown properties of space itself. For while the currently measured energy of empty space might one day disappear, who is to say there is not a much smaller residual energy of space that might remain? And who is to say whether this energy is positive or negative? If it is negative, the universe will ultimately collapse.

In 1999, my colleague Michael Turner and I showed that without an infinite amount of data about both the ultimate geometry of the universe and the ultimate vacuum energy of the universe, or a "theory of everything" which elucidates completely the structure of matter and forces in nature at a quantum level, including gravity, it is literally impossible to determine with absolute certainty the ultimate future of a currently expanding universe.

Since the former possibility would likely take an infinite time to amass and the latter possibility is, in my opinion, almost equally unlikely, this means that the ultimate future of our universe, and of time itself, may therefore forever be shrouded in mystery.

2

SPACE

Does space have an end?
Is there a smallest distance?
Are there hidden dimensions?
What is the geometry of space?
Is space fundamental or emergent?

Space is big. You just won't believe how vastly,
hugely, mind-bogglingly big it is. I mean, you may think
it's a long way down the road to the chemist,
but that's just peanuts to space.

—DOUGLAS ADAMS

Put three grains of sand inside a vast cathedral,
and the cathedral will be more closely packed with
sand than space is with stars.

—SIR JAMES JEANS

The way you have to relate to space
makes it like sculpture.

—ISABELLE HUPPERT

When I field questions about cosmology, I am often asked if the universe is infinite, and if it isn't, what is outside of it. They are easy questions to ask but not so easy to answer, not because we don't have good ideas, but because honest answers require more explanation than can easily be capsulized into a sound bite. So, here we go.

The second question has an answer that is at least easier to frame, if perhaps harder to picture. If the universe is finite, "outside" actually has no real meaning. Consider the simplest finite universe, a spherically symmetric "closed" universe—a three-dimensional geometry whose lower dimensional analogue, the two-dimensional surface of a sphere, is easier to picture.

Normally we think of such a surface as dividing space into two regions, that inside the sphere and that outside. But that is only because we are embedding the two-dimensional spherical surface into a space of one higher dimension. If we restrict ourselves to two dimensions, the surface of the sphere is all there is. There is no edge, no inside, no outside. If you travel on it long enough you will end up where you began.

Like a balloon, it can increase in size, but the surface is not expanding "into" anything. It is just expanding. If you cover the surface with small dots, as the surface increases in size each dot will move away from all the other dots, and no dots will move closer to any other dots.

If our universe is closed, it is just a one-dimension higher version of this spherical surface. It is curved in the sense that if you chose three perpendicular axes, labeled them x, y, and z in one place, and then followed the straight lines that defined them outward far enough in any direction, they would eventually be pointing in different directions than they were before. That defines curvature in three dimensions. It is not something we can directly visualize.

Similarly, a flat universe is not flat like a pancake, but rather defines the "sensible" space that we all imagine when we think of space—one in which all three perpendicular axes continue to point in the same direction everywhere.

There is one additional option for the geometry of space, if it is curved. It could have what is called negative curvature. In two dimensions, a negatively curved surface looks like a horse's saddle, except one that extends infinitely far in all directions. This type of geometry is called an open space.

Under normal circumstances, a closed space is finite in size, while flat spaces and open spaces are infinite. For these infinite spaces, while one doesn't have to worry about the conceptual question about what might be "outside" of them, the notion of an infinitely large space expanding does strain the imagination.

The simplest way to visualize that this is not a problem is to think locally rather than globally. Imagine an infinite flat space, for example, like an infinitely big rubber bedsheet. Now stretch it. Again, if one fills it with small dots, all the dots will move away from the nearby dots. The sheet will locally be "bigger," but the great thing about infinity is that it is infinite, and globally, the sheet will still be infinitely big.

If this bothers you, it probably should, but the bother is simply an inherent property of infinity, bringing to mind Woody Allen's line about eternity again. An expanding universe that is already infinitely big is another illustration of a property of infinite systems, a concept exposed most clearly by the brilliant German mathematician David Hilbert in what we often call Hilbert's hotel. Imagine a hotel with an infinite number of rooms, that I can label as one, two, three, four... and so on. Now say you try and check into this hotel, and the clerk tells you all the rooms are full. Just as you are about to leave, he says "Hold on! I can find you a room!" He simply puts the person in room number one into room number two, the person from room number two into room number three, the person from room number three into room number four, and so on down the line. Now all the rooms from two to infinity are filled, but room number one is now vacant and ready for your occupancy.

For much of the twentieth century, after Einstein developed his general relativity theory, the business of cosmology lay partially in trying to discover what the geometry of the universe was on its largest scale, namely whether we live in a flat or open universe, either of which would in principle be infinitely big, or whether we live in a closed universe, in which case space is finite in size, and if you look far enough out into any direction, you will eventually see the back of your head.

The reason for the intense interest in discovering the universe's geometry was that for a universe with gravitational dynamics governed by the energy of matter or radiation, geometry determines destiny. Namely, the ultimate future of the universe depends simply on whether the universe is open, closed, or flat, which is why determining which geometric pattern described our universe became the holy grail of cosmology.

It is also probably why I, as an elementary particle theorist, first became interested in cosmology. I reckoned that if I could determine how much dark matter existed in the universe, assuming it was made of a new type of elementary particle, I might be the first one to be able to pin down the geometry of our universe, and hence our future.

For some time, cosmologists recognized that exploring the nature of the mysterious dark matter that dominates the mass of our galaxy and essentially all observed galaxies—about which I shall have more to say in the next section—was essential if we were to determine the geometry of our cosmos. Studies of observable material—stars, gas, dust, and so on—had established by the early 1970s that they accounted for at most a few percent of the amount of matter necessary to close our universe, suggesting that the geometry of our universe was open. But this caused a theoretical problem.

For a universe dominated by matter or radiation, a flat universe—the borderline case between an open universe and a closed one—represented a point of universal instability. A flat universe was of course realizable, but a universe that was not exactly flat—either slightly open or slightly closed—would quickly evolve away from that configuration. It is analogous to balancing a pencil on its point. It is theoretically possible to do this, but the slightest draft in the room or vibration on the floor will quickly cause the pencil to topple over.

A universe like ours, which is more than ten billion years old, would have had plenty of time to evolve very far away from a flat appearance if it were not exactly flat. So even having a density of a small percentage of that needed to result in a flat universe today seemed implausible. Put another way, this would have required remarkable fine tuning, departing at very early times from exact flatness by less than one part in a billion, billion, billion, billion.

A simpler possibility suggested itself: the universe is, at any level of imaginably measurable accuracy, exactly flat today. While this on the surface might seem like an almost religious leap of faith, it turned out that inflationary cosmology, which by the mid 1980s had become accepted as the most likely picture describing the dynamics of the early universe, provided a natural "prediction" of a flat universe. Just as blowing up a balloon makes the two-dimensional surface of the balloon seem flatter at any point on the balloon, so the exponential expansion due to inflation would have driven the universe to become closer and closer to appearing flat as it became larger and larger. Even a modest period of inflationary expansion in the earliest moments of time would have driven the universe to have a density which would differ from the density of an exactly flat universe to an accuracy of better than hundred or even a thousand decimal places. The result of inflation would almost certainly be an essentially flat universe.

But there was a clear problem. Where was the remaining 98 percent or so of the material needed to produce a flat universe hiding?

Beginning with observations of the rotation rate of gas around our own galaxy and later confirmed by observing other galaxies and clusters of galaxies, the discovery that at least five to ten times as much mass exists in and around galaxies as can be accounted for by inventorying all the observable matter they contain presented an immediate potential solution to the problem. The universe could indeed be flat, and dark matter could make up the difference between the observed abundance of conventional material and the amount of matter needed to produce a flat universe.

This was a beautiful picture, and it led many physicists, myself included, to speculate on what the possible nature of dark matter might be, how we could calculate from first principles how much of it there was in the universe, and how we might directly or indirectly detect it and determine its makeup.

Beginning around 1990, however, another problem began to emerge. Careful studies of the gravitational dynamics of galaxies and clusters suggested that accounting for the dark matter dominating these systems would only produce about 20–30 percent of the total density needed to result in a flat universe. Once again, observers were being driven to the conclusion that the universe in which we live might be open, not flat.

In 1995, driven in part by the apparent contradiction between theory and observation, Michael Turner and I (and independently, a number of other theorists, including the cosmologist Jim Peebles) resurrected an earlier proposal we had made that the existing data in cosmology might only be consistent if we invoked a long dormant idea first proposed by Albert Einstein (and then almost as quickly discarded by him) that the dominant energy in the universe might reside in empty space—a possibility Einstein described as a cosmological constant.

This heretical proposal was really an action of last resort, and I, at least, made the proposal with the presumption that some of the observable data in cosmology must have been incorrect, because the notion that the expansion of the universe is dominated the energy of empty space, now called dark energy, seemed farfetched in the extreme. But, lo and behold, in 1999, two sets of observers who were measuring the expansion rate of the universe over cosmic time discovered, to all of our amazement, that the expansion of the universe was in fact speeding up—something that was only possible if the expansion was in fact being driven by the energy of empty space with the value that we had argued was necessary to produce consistency with the other data in cosmology.

This discovery sent shock waves through the scientific community and garnered the Nobel Prize for the observers in 2011. It also completely changed our picture, not just for the future of the universe, but also for the relationship between the geometry of space and that future.

Previously, as I have mentioned, there was thought to be a one-to-one correspondence between geometry and destiny. An open universe would expand at a finite rate forever, a closed universe would recollapse, and an exactly flat universe would expand ever more slowly but never quite stop.

Now suddenly, the discovery of energy in empty space changed everything. No matter the geometry of the universe, as long as it is expanding at the time it begins to be dominated by the energy of empty space, it will continue expanding forever…that is, unless the energy of empty space somehow disappears. Since, as I have also emphasized, the energy of empty space is perhaps the biggest known unknown in physics, as long as it remains a mystery, so does the future expansion history of the universe.

While the operational importance of understanding the geometry of space has therefore been somewhat diminished, the intrinsic importance of determining the geometry of the universe has not. Ultimately it would allow us to resolve one of the most fundamental known unknowns about our cosmos: is space finite or infinite?

Because inflation quickly drives the universe toward appearing flat today, and because the observable universe is measurably indistinguishable from a flat universe, at the level of better than about 1 percent or so, it may seem safe to presume that our universe is exactly flat. But, as I have earlier alluded, being very close to being flat and being exactly flat are two very different things. In fact, they are two infinitely different things.

Ignoring issues of topology, which I will touch on later, an exactly flat universe, like an open universe, is spatially infinite. But, like the earth, which may appear flat to an observer staring out at the horizon in Kansas, a closed universe can appear flat because inflation has blown it up to be so large that any signs of global curvature are pushed out well beyond the cosmic horizon. Both the earth and a closed universe are finite in spatial extension.

Unfortunately, unlike the earth, we cannot travel around our universe nor view it from the outside. As a result, the existence of a cosmic horizon beyond which we cannot see means that if we locally measure our observable universe to be flat, we cannot extrapolate this result to infer that the geometry of space outside the horizon remains flat. And the existence of dark energy makes the situation even worse. In this case, because of the accelerated expansion, the longer we wait, the less of space we will actually be able to observe. As I have mentioned, if we wait long enough—a few trillion years to be precise—then almost all distant galaxies will disappear beyond the horizon.

It is worth remembering however, that inflation is not a theory of the origin of the universe, just as Darwin's theory of natural selection is not a theory of the origin of life. Both instead describe phenomena that govern the subsequent evolution of their respective systems—space in one case and life in the other. When we go beyond inflation to consider the origin of space itself, there are good reasons to believe that the space that encompasses our observed universe must ultimately be closed.

As I have emphasized, once we consider the possibility that gravity is fundamentally a relativistic quantum theory, then the variables of that theory—namely, space and time—become relativistic quantum mechanical variables. And relativistic quantum variables can spontaneously be created and destroyed. Whole "virtual" spacetimes, like our own universe, could spontaneously pop into existence. But as I described at length in *A Universe from Nothing*, only those virtual spacetimes that have zero total energy might be expected to survive more than an instant, much less 13.8 billion years, before disappearing.

Energy is a well-defined concept in the static flat space we usually think of as the playing field for the events that we measure in everyday life. However, once spatial curvature and cosmic horizons begin to play a role, the definition of energy must be modified and can become more ambiguous. There is no well-defined, universally accepted understanding of the total energy of infinitely expanding flat or open universes. There is, however, an unambiguous understanding of the total energy of a closed universe. It must be zero.

The reasoning is relatively simple to state, even if it's somewhat hard to picture, and it is an extension of the statement that the total electric charge in a flat universe must be zero. In this case a simpler pictorial argument suffices. Ever since the work of the British physicist Michael Faraday, we can picture an electric charge as creating an electric field around itself, represented by radial lines pointing outward from the charge and extending out infinitely far. But, if the universe is closed, the curvature of space will be such that all those field lines will converge together again at a distant (antipodal) point. This will correspond to a charge of the opposite sign.

This can perhaps be more easily visualized if we consider one lower dimension, where it is easier to picture what is going on. On the surface of a two-dimensional sphere, if you draw lines emanating radially from a point, like the lines of longitude emanating from the North Pole, they will converge again at the South Pole. But if positive charges produce field lines that can be said to point outward, field lines converging inward on a point define a negative charge at that point. Therefore, it is inconsistent to imagine a single positive charge on such a surface. Effectively, it would always be accompanied by what you would measure as a negative charge in the opposite hemisphere. The same holds true for a closed three-dimensional space.

46

The situation is less easy to picture in the case of energy, but the fundamental idea is the same. Energy is the source of a gravitational field, and we can think of the flux of energy emerging from a region as defining the gravitational field throughout that region. In a closed spherical space, a source of energy in one place must be accompanied by a sink of energy at the spatial antipode. The total energy, like the total charge, of a closed universe must be zero.

Given this fact, it is easiest to assume that a universe that spontaneously pops into existence and survives long enough for life to evolve to the point of wondering about it is most likely to be a closed universe. It is also easier to picture a finite space popping into existence than an infinite space, though the dynamics of the universe are not determined by what is easiest for us to picture.

If we were placing bets on the ultimate geometry of our universe and the question of the finiteness of space, the smart money would probably bet that it is finite. But as both horse races and presidential elections have shown, the smart money can be wrong. Unfortunately, since observations may never allow us to measure the difference between a very slightly closed universe and an exactly flat universe, unless we ultimately develop a complete theory of the origin of the universe, we may never know the answer.

Mathematically there is a way to have one's cake and eat it too. Namely, one can have an exactly flat universe that is still finite in extent. I briefly alluded to this earlier when I used the word "topology."

If we take a flat piece of paper and identify two of the edges, we form a cylinder. If we identify the other two edges, we form a donut. Geometrically these look flat, in the sense that straight lines remain straight, but clearly unlike a flat piece of paper they have no edges, and unlike a flat piece of paper without edges, they are finite, not infinite.

It is theoretically possible that our three dimensions of space contain "non-trivial topology," namely, that distant points viewed in different directions are actually the same physical point, and our universe, though apparently flat, is finite in spatial extent. Indeed, one analysis of data by Roger Penrose and colleagues suggested evidence for non-trivial topology in our Universe. However, the analysis was subsequently disputed, and moreover, there is no compelling theoretical argument suggesting our universe would possess such features. As a result, this remains an unlikely theoretical possibility without current observational support. Nevertheless, as we open up new windows on the Universe, we could be surprised.

* * *

The question of the ultimate global structure of spacetime may be less directly relevant than the more immediate question of what exactly exists beyond our currently visible universe. On first glance, this too may seem like the kind of question that can have no answer, since we cannot directly detect anything beyond the reach of observation. Moreover, if the expansion of the universe continues to accelerate due to dark energy, that region will be forever inaccessible to us, hidden beyond our cosmic horizon.

Surprisingly, however, there is some hope of indirectly ascertaining what is beyond our horizon, at least if inflation is the proper description of the evolution of our early universe.

The rapid exponential expansion of space that takes place during inflation essentially wipes out all information of conditions before inflation began. However, quantum processes that occur during inflation leave remnant signatures in the universe today. One of these, the generation of small density inhomogeneities that can eventually result in the formation of galaxies and stars, is a central prediction of inflationary theories, and this prediction seems to be in good agreement with observations of the small temperature fluctuations in the Cosmic Microwave Background that would eventually evolve, through the coupling of radiation to matter, into galaxies today.

But mere consistency with observation is not proof. Other processes in the early universe could have created a similar spectrum of density fluctuations. There is, nevertheless, what appears to be a more unique smoking gun for inflation. Quantum fluctuations in matter and radiation during inflation produce the remnant density fluctuations just mentioned. But what about quantum fluctuations in gravity?

It is a characteristic of inflation that quantum fluctuations in all fields during inflation leave a remnant classical signal after inflation. For matter and radiation, this classical signal is manifest through density and temperature fluctuations measured in the Cosmic Microwave Background Radiation (CMBR). For the gravitational field, quantum fluctuations get turned into classical fluctuations in space and time. We would measure these today as a remnant background of gravitational waves.

Before turning to gravitational waves, I want to focus for a bit on the CMBR. I have mentioned this background radiation several times here but haven't really explained it. Nor have I described its significance for limiting how far out, and hence how far back in time, we can actually see in the universe.

As should be clear by now, the farther we look in the universe, the further back in time we are observing, because light takes a finite time to reach us. The furthest galaxies we could see with the Hubble Space Telescope take us back more than 90 percent of the way to the Big Bang, back to a time between about five hundred million and a billion years into the history of the universe, and we will see further back with the new James Webb Space Telescope.

We can't see back indefinitely, however, because when the universe was about three hundred thousand years old, with a temperature of about three thousand degrees, normal matter—made primarily of hydrogen atoms at that time—would have been ionized. At that time, every time a proton captured an electron to form a neutral hydrogen atom, energetic collisions with matter, or absorption of radiation, would have knocked out the electron. So, before that time, matter which now consists of neutral atoms consisted primarily of protons, electrons, and neutrons. But ionized matter is opaque to radiation because light can't directly escape from inside a dense gas of charged particles like protons and electrons.

Instead of traveling freely through space, light gets scattered, or in the language of quantum field theory, it gets absorbed and re-emitted, and so performs a random walk through the gas. Incidentally, this is also the reason that it takes almost a million years for the energy released in the nuclear reactions in the core of the sun to make it to the surface as light we can see. The photons perform a random walk of scattering, absorption, and re-emission as they proceed through the sun's interior. (I like to use this fact when I meet young-earth creationists. If nuclear processes have been powering the sun for only six thousand years, it wouldn't be shining as it does.)

In any case, once the temperature cools and protons capture electrons to form neutral atoms, the radiation that it had been in thermal equilibrium with would now be free to propagate unimpeded through the universe. This is the CMBR that our radiation detectors on earth and in space have measured. The radiation last interacted with matter almost 13.8 billion years ago, which is why it provides such an important probe of the early universe.

It also provides a sort of visual wall. As we look outward, we can see light from distant galaxies, quasars, and other objects that formed as structure grew in the universe. But if we try and observe the universe further out, to try and probe periods earlier than about three hundred and eighty thousand years after the Big Bang using electromagnetic radiation like light or radio waves, we cannot see past this "Surface of Last Scattering," as it is called.

While we cannot directly observe back to times earlier than the creation of this surface, we can hope to glean information about such times by observing that last scattering surface, which is what we are essentially doing when we detect the CMBR all around us today. The CMBR permeates all of space, and wherever you are when you detect it, you are detecting photons that effectively emerged from that distant spherical surface of space from which the photons were generated. In that sense, the last scattering surface is observer dependent. If the earth were located on the other side of our galaxy, the specific CMBR signal we might detect would have emerged from a different surface. The details might look different, but the statistical features would be the same. We can try to probe these features of the CMBR signal for some observable remnants of physical processes at much earlier times.

Let's return now to the gravitational wave signature from inflation. These waves are literally time-dependent oscillations in the fabric of spacetime. The Laser Interferometer Gravitational Wave (LIGO) observatory's discovery of gravitational waves from colliding black holes on September 14, 2015, was a monumental achievement in physics, worthy of the 2017 Nobel Prize that was awarded to its developers.

Because gravity is the weakest force in nature, it takes mammoth upheavals in the gravitational field to produce even a tiny signal in detectors on earth. In order to detect the turbulent collision of two massive black holes, each in excess of twenty times the mass of the sun, the LIGO detector had to detect a change in length between a laser and a reflecting mirror located four kilometers away by an amount less than one one-thousandth the size of a single proton over a second or two as the gravitational waves from that collision washed over the detector. As a theoretical physicist, I can honestly say that I thought the development of such a detector during the 1990s and early 2000s was a fool's errand. I say that even though I knew and admired the experimental and theoretical abilities of the physicists who designed and built it and had visited the laboratories where literally hundreds of dedicated physicists were working to make that discovery possible. Rarely have I been so happy to have been proved wrong as I was when I heard the first rumors of a detection early in 2016.

While the rapid expansion of spacetime during inflation produces gravitational waves on all frequencies, including those for which LIGO is designed, the magnitude of the inflationary signal is more than a million times smaller than the sensitivity of any existing or currently funded human-built detector.

However, the universe has provided us with a much more sensitive detector for very long wavelength gravitational waves. As I have already alluded, measurements of the CMBR have revolutionized cosmology and have turned it from an art into a high precision empirical science. In particular, the statistical features of the CMBR can reveal a wealth of information about the early history of the universe. The CMBR is a gift that keeps on giving.

In 1996, the American astrophysicist Marc Kamionkowski and his collaborators demonstrated that a careful statistical analysis of the polarization of the CMBR can provide a unique probe of any primordial background of gravitational waves.

Perhaps you wear sunglasses with polarized lenses. Light is an electromagnetic wave of oscillating electric and magnetic fields. If the electric fields in an oscillating light wave are all pointing in one direction, then the light is described as linearly polarized. Polarized sunglasses work because light that is reflected off water and other surfaces is often linearly polarized. Sunglasses that are designed to block out that light reduce glare from these reflections.

While microwaves are not visible, they too are electromagnetic waves, and they can be polarized. The CMBR radiation has many random polarizations, and it can also pick up additional polarizations when it scatters off dust in our galaxy on the way to our receivers. However, Kamionkowski and his collaborators demonstrated that long wavelength primordial gravitational waves present at the time the CMBR was generated can produce a helical snakelike trail of polarization in this radiation that can be discerned by carefully measuring the CMBR polarization at many points across the sky and doing a detailed statistical analysis of this data.

The problem is, the signal is very small. Temperature fluctuations in the CMBR are measured to be less than one part in ten thousand, and microwave technology that can reliably do this has been cutting-edge.

Because gravity is so weak, the polarization signal induced by gravitational waves from inflation is at best about one hundred times smaller than this. Nevertheless, brave experimenters have built detectors at various remote locations on earth and in space.

To get a sense of the challenge involved, an experiment called BICEP running at the South Pole sent shock waves throughout the physics community when a claimed discovery was announced, via a press conference and a published paper, of a polarization signal in the CMBR that precisely resembled that predicted by inflation. Moreover, it was at a level consistent with the maximum signal expected for inflationary models. The editors of the journal *Physical Review Letters*, which published the article, asked me to write an accompanying explanatory piece describing the significance of the result, which, had it been correct, would have been immense.

However, the much-heralded result was shortly thereafter shown to be consistent with polarization noise that could have been induced by scattering the CMBR with polarized dust particles in our galaxy. This realization wouldn't have been possible had it not been for the fact that a state-of-the-art space-based CMBR detector, the Planck detector, had just obtained a measurement of the dust contribution in the region of the galaxy corresponding to the direction probed by the BICEP detector.

As a result, while it was impossible to say that no part of the observed signal was due to inflation, it was also impossible to say with certainty that any part was. And, as Carl Sagan was fond of saying, extraordinary claims require extraordinary evidence.

Since the BICEP debacle, other groups, and the BICEP team as well, have developed more refined polarization detectors with sensitivities more than five times better than BICEP, and using Planck data, they have been able to choose directions in the sky where known dust contributions are small. No clear signal has yet been detected.

While this was bad news for the inflationary theorists who had hoped to ride this discovery to Stockholm, it should not be taken as a suggestion that inflation didn't happen. The BICEP claim literally corresponded to the maximum signal one might have expected from inflation, and most models predict a far smaller signal at or below the sensitivity of current detectors. So, it is still possible that a discovery might be made in the coming years.

I have dwelled on the possibility of discovering gravitational waves from inflation because of its potential significance for resolving the metaphysical mystery of a multiverse. A positive discovery could provide unambiguous evidence that inflation happened. Moreover, by measuring the detailed spectral characteristics of any such signal, one could get an empirical handle on just what type of inflationary model might produce such a signal. If we could so constrain the viable models, we could then ask ourselves whether these inflationary models result in the existence of a multiverse.

In this way, without ever being able to directly detect the existence of other causally disconnected spaces outside our observable universe, we could nevertheless develop extremely strong indirect evidence of their existence, turning metaphysics into physics.

Having only indirect evidence is a less than satisfactory way of accepting new realities, but it has a noble tradition in science. Consider atoms.

In 1905, in one of four landmark papers written within the span of months, Albert Einstein was able to calculate, on the basis of observations of the random motion of particles immersed in a liquid (so-called Brownian motion), that the liquid was made from individual atomic particles whose size he could estimate. Atomic theory had already laid the foundation of chemistry, but the reality of atoms was not fully accepted before 1905. Within a few years, no one doubted the reality of their existence. Indeed, a Nobel Prize was given for the 1909 experimental verification of Einstein's predictions. Yet for over four decades, until the development of sophisticated electron microscopes, there was no method to directly visualize atoms. The massive indirect evidence—beginning with Einstein, followed by Ernest Rutherford's scattering experiments to probe the physical makeup of atoms, and after that, by the use of X-Ray crystallography to scatter X-rays off atoms in crystals—left no doubt of their existence.

The observation of CMBR polarization due to a background of primordial gravitational waves might not provide indirect evidence at the level of precision of the atomic physics experiments, but it would nevertheless be a powerful step in providing a compelling case that our universe is not unique and that a multiverse exists.

Incidentally, the observation of gravitational waves from inflation would provide another striking bit of experimental data that would address a current known unknown that I will discuss in the next chapter. Classical gravity, described by general relativity, is inconsistent with quantum mechanics. Some physicists still wonder whether gravity can ever be described in terms of quantum mechanics or whether quantum mechanics itself will instead break down on the microscopic scales where quantum gravitational effects would otherwise become important.

Several years ago, my colleague Frank Wilczek and I showed that under very general conditions, the observation of gravitational waves from inflation would unambiguously require that gravity must be described by a quantum theory.

Unfortunately, as I have described, the experimental search for these waves is incredibly difficult, and this experimental difficulty is compounded by the fact that it is quite possible that inflation occurs at an energy scale small enough that the remnant gravitational wave background might never be observable. The ultimate evidence for the existence of a multiverse and gravity as a quantum theory may depend as much on luck as on scientific developments. Until then, both of these concepts must be considered just strongly motivated possibilities rather than empirically validated realities.

Even the possibility of a multiverse, however, provides incredible fodder for the imagination, even as it changes the very ground rules for physics. I became a physicist because I wanted to know why the universe is the way it is—namely, I wanted to understand what fundamental principles determined why nature acts the way it does. But if there is a multiverse, there may be no such fundamental principles. Many of the basic characteristics of our universe might be accidental. It could be that the laws of physics could be quite different in each universe within the multiverse, and that we measure the laws we do because our universe happened to allow the formation of galaxies, planets, and life. This idea, often going by the name anthropic principle, may seem distasteful, although my colleague Richard Dawkins thinks it has a certain quasi-Darwinian beauty. But, yet again, nature doesn't exist to please us, so like it or not, it may be true.

A more amusing possibility, if there is a multiverse, and if inflation is eternal, is that an infinite number of universes will ultimately be spawned. As I have stressed, infinity is very different than just "very big." If an infinite number of universes are formed over an infinite time, then the laws of probability suggest it is inevitable that there will be other universes just like our own, in which a planet like earth forms and life evolves, and everything we now see would exist as an exact copy, with perhaps a few exceptions. In that universe, the copy of you might be writing this book now, and the copy of me might one day be reading it, for example.

But even more dramatic options arise. There would be an infinite number of variations possible for those universes that produce earth-like planets, and perhaps even more for those that don't. For those that do, earth-like worlds with completely different histories could exist, as well as worlds with histories that are exactly the same except for one detail, and also others that are literally exactly the same. Every possibility that could exist, would exist. It is fodder for science fiction, perhaps, but for physicists, trying to imagine how to make quantitative predictions over a set of universes whose properties and probabilities we do not yet know is enough to give us headaches at best, or at worst to result in the occasional publication of articles in physics literature with mere metaphysical speculations.

On the other hand, it is of some solace that even as some universes may be in the final stages of their evolution, with stars dying out and any remnant forms of life disappearing, other universes are continually being born. Hope springs eternal in an eternally inflating multiverse.

In this sense, I view the possibility of a multiverse of nucleating universes embedded in an eternally expanding background of inflationary space as something akin to a modern workable form of Hoyle's steady-state universe. Recall that in this model, which died once evidence of the Big Bang became widely available and accepted, space was expanding but new matter was continually being created so that the density of the universe would always remain constant. In eternal inflation, on meta-scales far larger than those of any one causally connected universe, the multiverse always looks the same over time. New universes are being spawned, and outside of these, space is everywhere exponentially expanding, forever.

* * *

There is a German quote often attributed to Goethe, but which is apparently actually from Schiller, which can be liberally translated as: "The hardest thing of all to see is that which is in front of your eyes." Recent results in fundamental physics suggest that this may be literally as well as metaphorically true. Indeed, it is possible that whole new universes might exist just beyond the tip of your nose while remaining potentially eternally undetectable.

This hearkens once again to the famous wardrobe in *The Lion, the Witch, and the Wardrobe*, which I mentioned when talking about black holes, in which an entire new world was accessible by entering the wardrobe. I used this analogy in my book *Hiding in the Mirror* to discuss the long-time fixation with the possibility that extra hidden dimensions of space might exist beyond the three spatial dimensions of our universe.

Physics has not led to any real understanding of how to answer the question—one of the most fundamental questions about the universe one can ask, after all is, "Why is the space we inhabit three dimensional?" One possible answer is: maybe it isn't! On first thought, this answer seems ridiculous. We can explore space by moving around it, and I have yet to meet a (sane) person who has found a way to move beyond up/down, forward/backward, and left/right.

Nevertheless, the possibility that the world has more than three dimensions has fascinated artists, philosophers, and ultimately scientists for centuries. The physics motivation to consider extra dimensions began with independent musings by a mathematician and a physicist based on the similarities in form between electromagnetism and gravity.

Einstein revolutionized physics by describing gravity as a force that is associated with the curvature of spacetime. General relativity is a theory of the geometry of spacetime, which makes it fundamentally different than the other known forces in nature. Yet in its form here on earth, gravity looks almost identical to electromagnetism. Both forces fall off with the square of distance, for example. One has strength proportional to charge, and one to mass.

In fact, Maxwell's theory of electromagnetism can be cast in a form that looks a lot like a simplified version of general relativity. The electromagnetic field can be made to reveal a curvature just like the gravitational field, except that with electromagnetism, the curvature is not in real space, but rather in some internal mathematical "space." It turns out that this way of formulating electromagnetism is quite useful and important in physics, but for our purposes at the moment, we can think of it as simply a kind of mathematical trick.

After the development of general relativity, the Polish mathematician Theodor Kaluza, and independently the Swedish physicist Abraham Klein, wondered whether this mathematical trick might have a deeper significance. They reasoned that if there were an invisible extra dimension of space, perhaps electromagnetism would be associated with a curvature in this extra dimension, while gravity would reflect a curvature in the known four dimensions of space and time. It turned out that the mathematics worked out nicely, except for one thing. Unfortunately, unifying gravity and electromagnetism in this way would result in an additional force that is not observed to exist, which is one of the reasons why everyone has heard of Einstein, but many have not heard of either Kaluza or Klein.

In formulating their theory, Kaluza, being a mathematician, never bothered with the obvious question that Klein, a physicist, clearly needed to address. If there is an extra fifth dimension, why don't we see it? His answer was clever. If the extra dimension were curled up into a tiny circle, and were very small, none of the experiments we performed here on earth would be able to peer inside this very small circle.

There is a simple analogy that has often been used to describe this. Imagine a soda straw, with the length along the straw being space as we observe it, and the length around the circle enveloping the straw's circumference being the extra dimension. If that circumference gets smaller and smaller, eventually the straw will just look like a line to us, and the extra dimension will have become invisible.

This cute picture might have been consigned to the dustbin of history were it not for the development of what became known as string theory sixty years later. Physicists trying to develop a quantum theory of gravity—a theory which might consistently unify general relativity with quantum mechanics—discovered that a theory in which the fundamental spacetime objects were not points in space and time, but rather string-like objects, could be turned into a quantum theory, and the equations of general relativity naturally arise as the classical limit of this quantum theory.

This was a remarkable theoretical discovery, but it did not come without its own problems. The theory would indeed allow general relativity to be consistently quantized, but only if spacetime had not four dimensions, but twenty-six.

This was a lot to swallow, at least for many physicists. The mathematical beauty of the theory, however, caused a large cadre of very talented theorists to continue working on it, and they discovered that if one incorporated the existence of the other forces in nature, and the elementary particles that go along with them, it was possible to reduce the number of dimensions from twenty-six down to ten or eleven.

The details of how all of this came about are complicated, but happily not relevant for our consideration. I discussed them in *Hiding in the Mirror*, for those who are interested. What does matter, however, is the same question that Klein concerned himself with. If there are indeed other dimensions in nature, where are they hiding?

The answer proposed years later was the same one Klein had originally come up with. These extra dimensions could be curled up into a very small six- or seven-dimensional ball, and in this way could remain invisible. The diameter of the ball would be comparable to the scale at which quantum effects would become important in general relativity, about 10^{-33} centimeters, or about nineteen orders of magnitude smaller than the diameter of the nucleus of a hydrogen atom!

Needless to say, no existing or even proposed experiment could directly detect new physics, including the possible existence of new dimensions, on such a small scale—about fifteen orders of magnitude smaller than the scales explored at the most energetic particle accelerator now in existence, the Large Hadron Collider in Geneva, Switzerland.

For this reason—and the fact that the mathematics of string theory have become so complicated that the true nature of the theory, if there is one, remains elusive—string theory remains a fascinating area of study in mathematical physics, but whether it has anything to do with the real world remains an open question.

The possible existence of a tiny six- or seven-dimensional ball being attached to every point in our four-dimensional world, including at the tip of your nose, right in front of your eyes, may seem either romantic or idiotic, depending on your frame of mind. These extra dimensions, if they exist, remain impotent in the world of our experience and appear to be there simply to make the mathematics work out right. Moreover, there is no explanation whatsoever of why these extra dimensions should be curled up into small balls while our dimensions are potentially infinitely large. (Richard Feynman once harshly said about string theory: it doesn't explain anything—it just makes excuses.) And these extra dimensions would be too small to explore or visit, or for aliens to travel through on their way to visit us. No fun at all.

The story does become a bit more interesting, however.

In 1998, two different teams of researchers had a bright idea. What if forces like electromagnetism can't permeate extra dimensions beyond the four we know and love, but gravity can? That would help explain why these extra dimensions might be invisible to us, and it could also explain a mysterious feature of gravity that has long confused theorists: why is gravity so much weaker than the other forces of nature?

Gravitational forces, like electromagnetic forces, fall off with the square of distance in our three-dimensional space. But if there are more dimensions, gravity would fall off as a higher power of distance. If, say, the extra dimensions were not 10^{-33} centimeters in diameter, but 10^{-18}, then over the fifteen orders of magnitude of scale until one reached the size of the extra dimensions, gravity would fall off with a higher power of distance than would electromagnetism, which is a force that we posit cannot leak into the extra dimensions. On larger scales, gravity would begin to fall off with the square of distance as there would be no more extra room in the extra dimensions to leak into. This would mean that if we measure phenomena on scales larger than 10^{-18} centimeters, which is what we do with our particle accelerators and other laboratory experiments, gravity will behave like electromagnetism, but appear to be much weaker than it really is, having fallen off with a much higher power of distance on smaller scales than electromagnetism would.

This proposal had the potential to explain why gravity appears to be so weak compared to the other forces in nature on scales we can measure. The proposal also led to another interesting prediction. If the extra dimensions were as large as 10^{-18} centimeters then it would be possible to probe these extra dimensions with the world's most powerful particle accelerators today.

This was a cute idea, but having said this, it must be remembered that there was absolutely no explanation, within the context of this proposal, for why the extra dimensions should be large (or small, depending upon your point of view). Nevertheless, any time theorists make predictions that can be tested at accelerators, you can bet that the accelerator scientists will jump at the chance to rule them out. And they did.

Theorists can be very tricky however, and after this original proposal, a former graduate student of mine, Raman Sundrum, along with his collaborator Lisa Randall, and independently Savas Dimopoulos and colleagues, proposed that the extra dimensions could actually be infinite in size as long as gravity behaved rather peculiarly within them, and as long as gravity was the only force that could leak into extra dimensions.

Let me say at the outset that I found and still find the details of the proposal ugly, and I would bet good money that its content has nothing to do with reality. But aside from my doubts, it does open the romantic possibility that right under our noses could be a portal into huge extra dimensions, large enough not only to fit Narnia, but to fit whole new universes with exotic physics, and maybe galaxies and civilizations that we can never connect with. It is not in the least sense likely, but what remains surprising to me is that it is also not impossible.

Perhaps the most unusual proposal that has arisen out of string theory is that the very notion of spatial dimension itself may be misplaced. This proposal gets its name from the more well-established physics of holography, where full three-dimensional images are recorded on two-dimensional plates. If you look through these plates you can see a scene, and unlike a normal photograph, if you move your head, you can see objects emerging behind forefront objects.

In 1997, a young graduate student at Princeton, Juan Maldacena, made a bold conjecture: that the physics of a theory resembling the theory of the strong interaction between quarks, the particles that make up protons and neutrons in N-1 dimensions, was exactly the same as the physics of gravity and a particular type of curved space in N dimensions. This conjecture, called the AdS/CFT correspondence, if true, effectively means that all of the physics of a world governed by one very special type of interaction is identical to the physics of a world with a different number of dimensions governed by another very special, but different, interaction.

This conjecture, relating so-called conformal field theories, or CFT (the bread and butter of string theory), and anti-de Sitter (AdS) spaces (also an important type of space that crops up in string theories) in a different number of dimensions, is called the holographic principle. It put meat on an earlier proposal by Dutch physicist Gerard 't Hooft and was later promoted by American physicist Lenny Susskind.

Recall the information loss paradox associated with black holes. Gravity tells us that whatever falls inside the event horizon of a black hole falls into a space that is forever removed from the space outside of the black hole. Thus, the information about what has fallen in is forever lost. By considering quantum processes near the event horizon of a black hole, Stephen Hawking discovered that black holes should actually quantum mechanically radiate away their energy thermally, getting hotter the more radiation that has emitted and the smaller their resulting mass has become. The radiation that the black holes emit, being completely thermal, should carry away no information about what fell inside the black hole. Eventually, the black holes should radiate away entirely, leaving no trace of their existence or of any of the material that originally fell inside them. But quantum mechanics say that such information cannot be lost. One resolution is that all the information about what has fallen into a black hole is somehow stored in quantum correlations present just at the event horizon surface that may get transferred to the radiation, called Hawking radiation, that it emits. Like a hologram, the surface therefore stores all the information of the volume it surrounds.

While this idea remains interesting, it is not grounded firmly enough in a theory of quantum gravity for the physics community to accept it as a definitive resolution of the information loss paradox. Needless to say, Maldacena's conjecture, which does involve a potentially full theory of quantum gravity (albeit in a different number of dimensions than we experience), relating the surfaces which are the boundaries of spaces and the spaces of which they are boundaries, is highly suggestive.

Whether or not the AdS/CFT conjecture actually relates to the nature of our spacetime, as a tool it has been incredibly useful in physics because it allows us to relate the physics of some strongly interacting theories, which are themselves too complicated to perform accessible calculations with, to theories of pure gravity, which may be much easier to work with.

Aside from these technical issues, of which there are many, there is a deeper question here. If an N-1 dimensional surface can encode all the information associated with the N dimensional world it encloses, does dimensionality really mean anything significant? In this case, is the actual world N dimensional or N-1 dimensional?

Of course, we don't know. But for some, the enticing possibility remains that the four-dimensional world we call our universe is actually a hologram.

* * *

Is there a smallest distance? Is space, which appears on normal scales to be continuous, actually granular when we get down to what may be the smallest distance scales? Or is space instead an emergent phenomenon, arising as some effective approximation to something potentially far more exotic as we try to probe ever smaller distance scales?

These questions are at the heart of the effort to understand a quantum theory of gravity. And because we don't have an accepted theory of quantum gravity, the answer to all these questions is that we don't know, and we know we don't know.

Nevertheless, there are specific proposals that arise in the context of current theories, and I will discuss a few of these ideas here.

One of the reasons string theory is a successful potential theory of quantum gravity is not just that general relativity is naturally embedded in the theory, but also that there is one and only one fundamental parameter in the theory from which all other physical quantities can, in principle, be derived. This is the so-called string tension that determines in some sense the energetics of the fundamental objects in the theory: strings. In the original formulation of string theory, as I have described, the fundamental objects that comprise the playing field of physics were not points in space or time, but rather string-like excitations that swept out spacetime world lines. (In fact, as the theory has developed, the fundamental objects may ultimately not be strings, but rather things called branes, which are something like membranes, but may be higher-dimensional analogues).

In any case, inherent to the theory is the notion that at the scale where string excitations become important, a description of the universe in terms of points of space and time alone is no longer useful. More important, however, is a mathematical property of string theory called duality. Duality suggests, basically, that as one tries to mathematically describe the dynamics of the universe on scales smaller than the string scale, the description becomes mathematically equivalent to examining physics on scales larger than the string scale. Effectively this implies the string scale is really the minimum physical scale that makes sense. It simply doesn't make sense to consider the dynamics of possible excitations on smaller scales.

The string scale, set by the string tension, is generally set to be the scale where quantum effects in general relativity can no longer be ignored. This scale, called the Planck scale, is around 10^{-33} centimeters, as I described earlier. In this sense, string theory suggests that space as we normally understand it can only be described on larger scales.

Fixing a minimum distance scale for the universe is, again roughly speaking, one of the reasons that string theory fixes the problems of quantum gravity. In general relativity, quantum effects become ever larger on every smaller scales. If one extrapolates to zero distance, these effects become infinitely large. But if there is no zero distance, there are no infinities.

I consider string theory to be the leading candidate for a theory of quantum gravity, although that does not mean I think it is likely to be the correct theory. I remain somewhat agnostic on this issue. It is just that string theory is the most natural extension of the mathematics that has successfully allowed us to describe all the other known forces in nature.

In any case, there are other ideas floating around. A competing idea to quantize gravity is something called loop quantum gravity. Less well-motivated, in my opinion, this theory nevertheless has dedicated adherents who have been working hard to get new and interesting results over the years.

As the name suggests, the fundamental objects in loop quantum gravity are loops of spacetime. These loops weave together, again on the Planck scale, into a network of loops from which spacetime emerges. Once again, in this theory, it makes no sense to talk about spatial distances smaller than the Planck scale because space itself is not defined on smaller scales.

This network of loops is reminiscent of an earlier idea promoted by the iconic theoretical physicist John Wheeler (of black hole fame), who argued that space on its smallest scales should be thought of as a foam-like structure, which he called spacetime foam, due to quantum effects.

There are other, even less well-formulated (again, in my opinion) ideas that literally say that space on its smallest scales is a discrete structure, and that space emerges as discrete structures, like a lattice, which become invisible so that space seems continuous, just as, say, a diamond seems continuous even though it is made of a lattice of carbon atoms at its smallest scales. Note that below the scale of a few carbon atoms the notion of "diamond" becomes ill-defined.

Perhaps the most recently promoted idea, which has gained attention outside the physics community, is the cyclic conformal cosmology proposal of Roger Penrose that I mentioned earlier. His proposal has gained little traction inside the physics community, but it is reminiscent of an idea I am sure many of us (and some science fiction writers) have had, and which achieved its most amusing form in the film *Men in Black*. There, entire galaxies with numerous civilizations exist as small objects within a marble in a world existing on much larger scales.

Penrose argues that at late times in the evolution of our universe, after matter like stars and galaxies have largely disappeared into black holes, the universe will be such that length itself loses its meaning. Lengths, which for the long dead observers who lived on the stars and galaxies separated by vast expanses of space, could then, he argues, be identified with lengths much smaller than those in a subsequent, later universe would now call the Planck length. In short, the birth of new universe could emerge out of the death of an old one, with all the vast expanse of space between dying black holes in the old universe too small for the observers who evolve in the new universe to even measure. In doing so, and by requiring some new unusual physics, Penrose argues that the new universe could avoid the nasty singularities of a Big Bang and could naturally have the particular features our current universe seems to have. Sound crazy? I think it probably is. But sometimes crazy ideas are true. Just not very often...

3

MATTER

What is the world made of?
How many forces are there?
Is anything fundamental?
Is quantum mechanics true?
Will physics have an end?
Will matter end?

You may not feel outstandingly robust, but if you are an average-sized adult you will contain within your modest frame no less than 7 x 10^{18} joules of potential energy—enough to explode with the force of thirty very large hydrogen bombs, assuming you knew how to liberate it and really wished to make a point.

—BILL BRYSON

Dirt is not dirt, but only matter in the wrong place.

—HENRY JOHN TEMPLE

Matter is plastic in the face of Mind.

—PHILIP K. DICK

The things that matter most are not things.

—ART BUCHWALD

n 1936, after the discovery of the muon, a heavy cousin of the electron, in cosmic ray showers, the Nobel Laureate and experimental physicist, I. I. Rabi, famously quipped "Who ordered that?" We are still asking the same question.

Earlier, I described the acceptance of the atomic picture of matter by about 1905, in spite of the fact that atoms could not be directly seen. At the same time, given how our lives are now governed more or less completely by physics at the atomic scale, it is amazing that just a little over a century ago atoms still were often viewed merely as hypothetical constructs that allowed easy classification of materials and their chemical interactions.

It was almost at the same time, before the discovery of relativity, atoms, and quantum mechanics, that the famous Scottish physicist Lord Kelvin was attributed as saying there was nothing new to be discovered in physics and all that remained was more and more precise measurement. In 1894, the great American experimental physicist Albert Michelson (I was honored to be appointed to the same Chair of Physics that he held at the same university, a century later) was more explicit:

> "...it seems probable that most of the grand underlying principles have been firmly established and that further advances are to be sought chiefly in the rigorous application of these principles to all the phenomena that come under our notice. An eminent physicist has remarked that the future truths of Physical Science are to be looked for in the sixth place of decimals."

It is easy to understand this hubris. Two hundred years earlier, Newton had developed a theory of gravity that explained the motion of everything from cannonballs to the planets around the sun. A generation prior to the statement, a complete and elegant theory of electromagnetism, the only other force known in nature at the time and the one that appeared to govern the properties of materials on earth, was developed by the Scottish physicist James Clerk Maxwell. The dynamics of everything that could be seen was apparently understood.

The problem of course is that there is a lot more to the universe than can be easily seen with the naked eye or even with a microscope. In the hundred or so years since Michelson's statement was made, we have discovered literally a whole new universe on scales smaller than the size of atoms, two new forces, and a host of new elementary particles.

We now have a theory, the so-called Standard Model of particle physics, that appears to correctly predict the results of every experiment we have performed at subatomic scales. Yet it is interesting that one doesn't see the same parade of eminent physicists today saying that everything is understood up to the first six decimal places.

The reasons for this are threefold. I will outline them first, and the rest of this chapter will focus on the details.

First, the Standard Model, as impressive as it may be, is notoriously incomplete. It has, depending on how one counts, at least eighteen free parameters that have no underlying explanation and must simply be fit to data. It is worth emphasizing that one of these parameters is perhaps the most important parameter of the entire theory: the basic energy scale at which two of the three forces described by the theory get unified into a single theory. Moreover, not only is this scale essentially ad hoc, but our basic ideas about quantum physics also tell us that the value it has—some seventeen orders of magnitude smaller than the energy scale at which quantum gravitational effects become important—is unnaturally small. Without some new, beyond the Standard Model mechanism to stabilize this value, this scale and the quantum gravity scale should be roughly equal.

Second, unlike the situation in 1900, there are several known notorious and outstanding mysteries about nature that beg for an explanation the Standard Model does not provide. These include two cosmological conundrums: (a) the fact that the dominant form of mass that appears to govern the dynamics of galaxies and clusters of galaxies apparently cannot be composed of the same fundamental constituents that make up visible matter—protons and neutrons—and is therefore most likely made up of some new form of elementary particle that is not part of the Standard Model; and (b) the even stranger fact that the dominant form of energy in the universe doesn't seem to correspond to any matter at all, or even any form of radiation. Rather, as we have seen, this energy appears to reside in empty space itself, and this has completely defied any attempts to understand it to date. It is perhaps the biggest outstanding mystery in cosmology and fundamental physics.

Alongside these cosmological mysteries is a clear mystery from particle physics. My favorite elementary particles in nature, neutrinos—ethereal particles released during nuclear reactions, whose interactions with normal matter are so weak that neutrinos streaming at us from the nuclear reactions inside the sun can go right through the earth without a single interaction—have a mass that is not predicted in the Standard Model, or even easily accommodated within it. Some new physics must be involved. We just don't know what.

Finally, there's gravity. Gravity stands out among the four known forces in nature because classical general relativity is incompatible with quantum mechanics. Something has to give—either gravity or quantum mechanics. Either way, something dramatically new is required.

* * *

Humphrey Bogart, playing Sam Spade in *The Maltese Falcon*, made movie history with his reworking of a line from Shakespeare when he called the infamous eponymous artifact "the stuff that dreams are made of." Over the millennia, as science has tried to come to grips with the fundamental nature of matter, another reworking seems appropriate. From the "atoms" of Democritus to the quarks of Gell-Mann, philosophers and scientists have engaged in an exercise that might be aptly described as "the dreams that stuff is made of."

Over and over again, abstractions proposed as mere placeholders to aid in helping parse the stupefyingly diverse complexity observed in the world around us have turned out to be real, though usually in surprising ways.

Democritus, in the 5th century BC, first speculated that matter might be composed of individual atoms, without offering any concrete understanding or explanation of what these atoms might be like. When atoms began to be understood as real entities, the first person to directly probe them, Ernest Rutherford, discovered something totally unexpected.

By this time, it was recognized that atoms were not indivisible entities. The first subatomic particle, the electron, had been discovered only a decade earlier in 1897, by the British physicist J. J. Thomson. Thomson, like others, had been searching for the source of charge in electric currents, and he identified the electron as a unique particle in cathode ray tubes subjected to a magnetic field, allowing him to measure the charge to mass ratio of the electron, which he found to be dramatically different than the previously measured charge to mass ratio of atoms. Assuming equal charges, the mass of the electron was determined to be about two thousand times smaller than the mass of a hydrogen atom. This crushed the notion that atoms were the smallest fundamental particles.

Following that, it was assumed that atoms were made of some uniform heavy material with electrons embedded inside, like raisins in a pudding. But when Rutherford bombarded gold atoms with other atomic mass-scale objects, he discovered something astounding. Most of the time the projectiles would not be deflected at all, but every now and then they would be deflected right back toward the source. This indicated that atoms were mostly empty space, with something incredibly heavy and dense at their centers.

We now realize that these central objects are atomic nuclei, but at the time, the composition of these heavy centers was not known. Rutherford determined the mass of the lightest nucleus, hydrogen, which he realized was a candidate for a new elementary particle, which he labeled the proton, from the Greek for "first."

For some time, it was believed that atoms must consist of a nucleus containing protons surrounded by a cloud of electrons. The problem was that the mass of heavier atoms was larger than could be accounted for if there were an equal number of protons in the nucleus as electrons surrounding the nucleus. One possibility was that there were more protons inside the nucleus, but that some electrons also resided in the nucleus so that atoms would still remain overall electrically neutral.

Ultimately this conundrum was resolved in 1932 when James Chadwick discovered a new, neutral subatomic particle: the neutron. With this discovery, atomic nuclei were understood to contain both protons and neutrons, accounting for the mass of heavier atomic nuclei while allowing nuclei to have a number of protons equal to the number of electrons surrounding the nucleus, so that atoms could be electrically neutral.

This could have been the end of the story, except that neutrons, which are actually the most abundant particles in all matter heavier than hydrogen (since heavy nuclei generally have more neutrons than protons) are actually radioactive.

A free neutron, not confined within an atomic nucleus, has a half-life of about ten minutes, meaning that if one has a pile of free neutrons, half of them will decay within ten minutes.

We will get to their decay products in a bit. For the moment though, let's concentrate on the decay itself, which is remarkable since, as I have indicated, there are more neutrons in your body than any other particles, and most people survive a lot longer than ten minutes.

The resolution of this apparent paradox comes from the fact that the neutron and proton have almost exactly the same mass. The neutron weighs only about one part in a thousand more than the proton. This difference is enough to allow free neutrons to decay into protons (and other light particles). However, when a neutron is bound into an atomic nucleus, it loses energy as it is captured. Since Einstein taught us there is an equivalence of mass and energy, when neutrons are bound in atomic nuclei, they actually lose enough mass that energetically, they can no longer decay into protons, a remarkable accident of nature that makes stable matter stable.

There remains a mystery associated with the decay of neutrons that persists today. There are two different ways of measuring the lifetime of the neutron. One method, involving a neutron "beam," measures the number of neutrons in the beam entering a region and the number leaving the region some distance (and time) later. The other method involves trapping neutrons in a magnetic bottle. While neutrons do not have an electric charge and therefore cannot be manipulated by charged electrodes, neutrons nevertheless act like little magnets, and therefore magnetic fields can be used to produce a force that keeps them away from the walls of a container, more or less at rest. When a neutron decays in the trap, a proton and an energetic electron are emitted. By measuring these decay products, we can attempt to measure the neutron's lifetime.

And here's the rub. Both methods should yield the same number, but they don't. And ever more sensitive experiments using both techniques keep giving different numbers. The lifetime in one differs from the other by about five seconds or so.

Five seconds doesn't seem like a lot, but when the experimental sensitivity in each technique is said to have an uncertainty of less than two seconds or so, this difference is uncomfortable, if not yet definitive.

Whenever one is presented with a problem like this, there are two solutions. Either one technique or the other is flawed and the apparent difference is not significant, or there really is an underlying physics reason for the difference. One group of individuals has claimed that perhaps there is another decay mechanism for neutrons that doesn't emit protons and electrons, but rather particles of dark matter. These particles would be invisible in the decay-counting experiments and would suggest that fewer neutrons are decaying than actually are—hence the longer lifetime.

It is too early to know if new physics is required. I am betting not, because in my lifetime these kinds of anomalies have almost always ended up being detector-related. But I would love to be wrong. At this point, we don't know.

Neutron decay produced another puzzle that sparked physicists to make an outlandish proposal about matter that turned out to be true, and then years later, they were flabbergasted to find that their proposal, far from being too crazy, wasn't crazy enough.

I have described how neutrons were observed to decay into protons and electrons. However, a problem quickly arose. When a particle at rest decays into two particles, having mass of, say, m1 and m2, then a fundamental law of physics, the conservation of momentum, says that the two outgoing particles must be emitted in equal and opposite directions, and their velocities will be fixed so that the outgoing velocity of particle one is equal in magnitude to the velocity of particle two divided by the ratio of their masses.

This is just to say that the two particles have fixed outgoing velocities that cannot vary. But since these velocities also determine the energy carried by these particles, that means the outgoing particles will be measured to have fixed energies. (In the case of one of the two outgoing particles being much heavier than the other—as in the case of neutrons decaying into protons and electrons, with the electron's mass being one two-thousandth of the mass of the proton—essentially all of the available energy in the decay will be carried away by the electron.)

But when experimentalists measured the energy of the outgoing electrons in neutron decay, they discovered that these electrons could be emitted with a wide variety of different energies, violating these basic restrictions coming from the two pillars at the heart of classical physics, the conservation of momentum and of energy. The towering theoretical physicist of the time, Niels Bohr, was driven by this conundrum to suggest that perhaps at subatomic levels one might have to give up these sacred conservation laws.

Craftier still was the Swiss physicist Wolfgang Pauli, who realized that the problem could be solved if a third, invisible particle was emitted in neutron decay. This particle could take up the momentum and energy not carried by the electron. Pauli proposed this particle in a joking letter to eminent colleagues Lise Meitner and Hans Geiger.

This suggestion was outrageous for a variety of reasons. First, the new particle had to be very light—much lighter than even the electron because there was so little available energy in those decays after the energies of the electron and proton were taken into account. Second, it had to be effectively invisible, since it was not observed in experiments. This meant it not only had to be electrically neutral, but it also had to interact far more weakly with matter than any particle yet known in nature.

The great Italian physicist Enrico Fermi was not daunted. He thought Pauli's idea was beautiful, and he even gave a name to the new hypothetical particle: the neutrino, meaning "little neutral one." He included it in a new theory of neutron decay that would eventually help form a key element of the Standard Model of particle physics.

Needless to say, twenty-three years later in 1956, an equally inventive experimental physicist, Fred Reines, and his collaborator Clyde Cowan, figured out a way to detect this elusive particle by observing a handful of neutrino interactions from among the billions and billions of neutrinos emitted in the nuclear decays in nuclear reactors. The imaginary neutrino was real.

As I have already stated, neutrinos are my favorite particles in nature. The reason is that, beyond their otherworldly characteristics, since their discovery they have been involved, in one way or another, in almost every major discovery about the nature of matter and the forces that govern the dynamics of matter. I'll explain some below.

The first point involves neutron decay again, as well as neutrinos themselves. Many elementary particles, including the particles that make up atoms—electrons, protons, and neutrons—behave as if they are spinning. That is, they carry angular momentum with them, like a spinning top. Due to the vicissitudes of quantum mechanics that we will discuss shortly, they are not actually spinning like tops, but are nevertheless measured to have non-zero angular momentum, as if they are spinning around some axis. It turns out in the quantum world, they can be spinning around all axes at once until we measure them to have angular momentum around some specific axis.

Electrons, protons, and neutrons have what is called spin ½, which means along any axis we choose to measure, they can have two different spin states. They can either have +½ a unit of angular momentum or −½ a unit of angular momentum about that axis, or they can be in some linear combination of the two different spin states.

If a particle like an electron is moving in some direction, that direction defines an axis, and the spin angular momentum of the particle can either point forward along the axis, backward along the axis, or be in some linear combination of these two states. If the spin angular momentum points forward along the axis, we say the particle is right-handed. If it points in the opposite direction, we say it is left-handed.

The reason for this nomenclature is relatively simple to explain. Consider a top that is spinning clockwise along some axis as it is traveling along in the same direction. In this case, the spin angular momentum of the top turns out to point in the direction of motion.

Now, say that direction is toward a mirror. In the reflection in the mirror, the top will appear to be traveling in the opposite direction (outward toward you), but it would still appear to be spinning clockwise. So, in the reflection in the mirror, the top appears to have spin angular momentum pointing in the opposite direction of its motion.

Now take your right hand and curl the fingers around while pointing your thumb outward away from you. Notice that your fingers are curled up clockwise. Thus, the thumb on your right-hand points in the direction of the spin angular momentum of a top spinning around in the direction represented by your curled up fingers. If you look at your hand in the mirror, with your thumb pointing toward the mirror, your right hand will now look like a left hand, with the fingers curled counterclockwise compared to the direction your mirror-thumb is pointing in. The thumb on your left hand therefore would point in the direction opposite the spin angular momentum of a top spinning around in the direction of your curled fingers.

A second interesting story about neutrinos occurred in the same year that Fred Reines discovered they were real. In the summer of 1956, two young particle theorists, T. D. Lee and C. N. Yang, were spending the summer at the Brookhaven National Laboratory, a particle accelerator facility. At that time the physics community was perplexed by a mystery. Two seemingly different particles had precisely the same masses and lifetimes but decayed into different sets of decay products. That was very strange. It might have been tempting to assume these two different particles were really the same particle that simply had two different decay modes, with some particles decaying into one set of decay products and some particles decaying into others. That happens all the time.

The problem was that in this case it seemed impossible.

A fundamental law of nature is called parity conservation, meaning that the world would look the same in a mirror if left were interchanged with right. It seems eminently reasonable to assume that physical processes shouldn't be able to distinguish between left and right, because these are human inventions, after all. Some of us have trouble remembering left versus right precisely because they are arbitrary labels.

That is not to say that all objects are left-right symmetric. Your face, for example, is not perfectly symmetric. Parity conservation just says that when you look in the mirror, you shouldn't see anything strange just because the left and right side of your face are now interchanged.

Combinations of particles can have either "odd" parity or "even" parity. That means if look at the combinations in the mirror they will either be identical, or they will be like your hand, with right turning into left. For example, if I had two electrons, one left-handed and one right-handed, and I looked at them in the mirror, one would still be left-handed and one right-handed. The identities of the electrons would have switched, but since electrons are all identical, the combined system would not have changed compared to the original system.

If parity is conserved, a system which initially has even parity cannot evolve into a system with odd parity, even if one of the particles in the system decays.

Now, returning to the puzzle of the decaying particle in 1956, it turns out that the two different configurations of decay products of the two particles in question had opposite parity. Thus, it was impossible that the two original particles could represent the same type of particle, since its parity had to be either odd or even.

Lee and Yang instead took an agnostic approach. Rather than have faith that parity was conserved in all particle interactions, as common sense dictated, they decided to see if there was any experiment that had ever been done that could test if parity *wasn't* conserved. Because if it wasn't conserved, then the decay puzzle of those two particles, called the tau and theta particles, could be solved. They could both be the same particle, as one would otherwise have suspected if not for the parity measurement of the decay products.

Sure enough, they discovered that there had been no such experiment performed that would have tested whether parity was conserved in these types of decays. Lee and Yang suggested several experiments, and the simplest one involved neutron decay, usually called beta decay, because electrons, originally called beta particles, are emitted.

If a neutron was prepared using magnetic fields in an initial configuration with the spin of the neutron pointing up, then if parity were conserved, one would expect to see as many decays in which the outgoing electron traveled in the upward hemisphere as the downward hemisphere. Otherwise, if we called up "left" and down "right," nature could distinguish between the two. If parity were violated, one would expect to see more events recorded in one hemisphere than the other.

Almost immediately after writing their paper suggesting the experiments, two such experiments were performed, one involving neutrons decaying and another involving a very similar decay of muons (those heavy electrons) into electrons and neutrinos. Both experiments demonstrated definitively that not only was parity not conserved in those decays—mediated by a force we now call the weak force—but it was maximally violated. In advance of the experiment, the ever-skeptical joker Wolfgang Pauli said in another letter to a colleague that he couldn't believe that God was a weak left-hander. (I am assuming, given that he was Swiss, that he wasn't using baseball terminology.) He was delighted to be proven wrong.

Physicists had discovered that, as far as the weak interaction, or the force mediating the decays of these particles, is concerned, nature *can* tell the difference between right and left!

So, where do neutrinos fit into all of this? Well, the ultimate proof that parity is maximally violated by the weak interaction becomes manifest when examining the properties of the only particles in nature that just feel the weak interaction and not the electromagnetic interaction or the so-called "strong" interaction that governs the dynamics of protons and neutrons. Namely, the neutrino. In nature, as far as we can tell, the neutrinos we observe are *only* left-handed. They always spin in the opposite direction of their motion. They are the only particles with this property—and that is what distinguishes us from a universe in the mirror, which would only contain right-handed neutrinos.

More generally, for particles like electrons, which can be either left-handed or right-handed and which are sensitive to both electromagnetism and the weak force, the left-handed versions interact differently when being acted on by the weak force than their right-handed counterparts.

Why does the weak interaction have this property? We don't know, although we expect that the answer will lie in understanding the relationship between the weak force and the other forces in nature.

That relationship begins with a remarkable connection between the weak force and electromagnetism. On the surface, these forces seem quite different. One force is long-range, the other operates only on scales smaller than an atomic nucleus. One is strong enough to govern all of chemistry, while the weak force is so weak that neutrinos emitted from nuclear reactions can traverse the whole earth without a single interaction.

It was hypothesized in the 1960s that the reason the weak interaction seemed so different from electromagnetism could be explained if the particles conveying the weak force were heavy. In the quantum theory of electromagnetism, the electromagnetic force is conveyed by an exchange of photons, the quanta of the electromagnetic field, and the particles that make up visible light, radio waves, X-rays, and more. Photons are massless, and as we know, Einstein famously demonstrated that they travel at precisely the speed of light.

The reason the electromagnetic force is long-range is directly related to the massless nature of photons, which, because they are massless, can be exchanged between particles at long distances with vanishingly small amounts of energy being involved.

If the weak force was similarly attributable to the exchange of particles, its short-range nature and its weakness could be explained if the force carriers were not massless, but very heavy, almost one hundred times heavier than the mass of the proton. Such heavy particles cannot be easily exchanged at large distances because they carry a large amount of energy due to their rest mass.

An elegant mathematical framework for how this might work was first proposed by Sheldon Glashow in 1961. In his picture, photons would be exchanged between particles in a manner more or less identical to the exchange of these new force carriers, which we now call the W and Z particles, except that the latter would be extremely heavy, and the couplings between the W and Z particles and other elementary particles would be different for left- and right-handed particles. In this way, the two seemingly different forces could be unified in a single mathematical framework.

This was mathematically valid, but it begged a lot of physical questions. The elephant in the room remained the elephantine mass scale of the W and Z particles. Why would they be heavy and the photon remain massless? The answer to that question relied on the development of a remarkable idea that has since become a central part of the Standard Model of particle physics, of which Glashow's initial tentative proposal became a nascent part.

In 1967, Steven Weinberg and independently, Abdus Salam, proposed that a remarkable mechanism developed two years earlier might provide the key. This mechanism was developed by several groups but is now most closely associated with Peter Higgs. At the heart of the proposal was the idea that all elementary particles, including the then proposed W and Z particles, were fundamentally massless and that the masses we measure them to have been accidents of our circumstances.

This was a remarkable claim, and it surely required remarkable evidence, which it took almost fifty years to obtain. The gist of the idea is that there is a field permeating all of space, and that field is associated with a new elementary particle called the Higgs particle or Higgs boson. Elementary particles that interact with this field experience some resistance as they move, like pushing a shovel through mud, and this inertia results in their behaving as if they are massive. Particles that interact more strongly with this field appear heavier, and those that interact less strongly, lighter. Some particles, like photons, don't interact with the field at all and remain massless.

As fantastical as this idea is, what made it so seductively attractive to theorists was that it provided a mathematical mechanism to give the W and Z particles mass without destroying the beautiful underlying mathematical symmetry, called gauge symmetry, that otherwise connected the theory describing the weak interaction with that describing electromagnetism. Any other way of imposing a mass on the W and Z particles would have explicitly destroyed this connection and blocked any attempt to develop a theory of the weak interaction that might produce sensible, calculable, results.

In physics, of course, the proof is in the testing, not in the beauty, regardless of what some people might whimsically wish. The first step in testing was to see if the infamous W and Z particles existed.

The only way to do this was to build a machine powerful enough to produce these particles and precise enough to find a needle in a haystack. The first requirement was demanding, as it required a new high energy accelerator to be built. The second requirement was even more daunting since the W and Z are involved in weak interactions, and if particles like protons collide together at high energy, there are billions more events produced due to the interactions of these strongly interacting particles compared to the few interactions in which a W or Z particle might be directly produced.

Amazingly, in spite of the difficulties, in 1983, detectors at an accelerator at the European Center for Nuclear Research (CERN) outside Geneva recorded the discovery of both of these particles. Once again, neutrinos played a role.

In order to untangle the rare W events from the huge background, a very special signal had to exist. W particles can decay into an electron and a neutrino (actually an antineutrino, but for the purposes of this argument they can be considered essentially identical). In the detectors, the electron would leave a visible charged track, but the neutrino would escape undetected. The signal of W production would then be the observation of a single high energy electron with no other particles coming out opposite to the electron to balance its momentum.

No other process could produce such a unique and unusual signal, like the sound of one hand clapping, as the famous Zen koan might say. Or, perhaps another literary allusion, this time from Sir Arthur Conan Doyle, is more apt: it would be as significant a clue to unraveling the mystery of the weak interaction as is the dog that doesn't bark in the night.

On January 21, 1983, at a seminar at CERN, the experiment's leader, Carlo Rubbia, presented six clear events with this signature out of the billions analyzed by one of the two detectors searching for the W and Z. Shortly thereafter, a similar number of other events unveiled the reality of the Z. The carriers of the weak force, the heart of the Standard Model, had been revealed exactly where they were predicted to be.

As important as this experimental discovery was, one central— and seemingly, the most contrived—piece remained missing from the elegant electroweak unification theory. The piece that gave the W, Z, and all other elementary particles mass: the Higgs field.

I say it seems contrived because while the W and Z particles seemed rather natural extensions of electromagnetism, resembling heavy copies of the photon, the Higgs field required the existence of a new type of elementary particle—not surprisingly, called the Higgs particle (mentioned previously)—which interacted with itself and other particles in a way that could produce the required effect, but which was not suggested or required by any other basic principles of physics. I know that I remained very skeptical that nature would have actually adopted this specific mechanism to achieve the goal of giving particles mass, and I felt that the Higgs was, once again, just a convenient mathematical kluge that was a sort of placeholder for other, and perhaps more interesting, physics. I remember the statement by Sheldon Glashow that the Higgs mechanism was like the "toilet" of the Standard Model, where things one didn't want to think about were hidden.

For twenty-nine years after the discovery of the W and Z, experimental physicists threw everything they had into the effort to discover the Higgs mechanism or rule it out. Because its mass wasn't strongly constrained in the theory, as each new machine with ever higher energy turned on, Higgs hunting became an activity. Ultimately, the particle physics community decided to build a machine guaranteed to find the Higgs or rule it out. That machine, the Superconducting Super Collider (SSC) was never built, largely due to political issues in the US. Instead, an existing accelerator at CERN was upgraded to become the Large Hadron Collider. Cycling two beams of high energy protons in opposite directions around a 26.7-kilometer tunnel, each proton was accelerated to have an energy of over six thousand times its rest energy before colliding with a proton traveling in the opposite direction.

The discovery of the Higgs, on July 4, 2012, was fascinating. I was traveling between Australia and the US during the months before this, talking to experimentalists involved in the CERN experiment from both sides of the Pacific. In April, I learned from one experimentalist that they had ruled out the entire allowed mass range for the Higgs except for a narrow window around 125 times the mass of the proton, and they expected to be able to rule out the remaining range as new data was analyzed.

As someone who was dubious about the Higgs, I found this quite exciting, because if the Higgs wasn't there, then something else even stranger must be going on.

But nature had other ideas. In that narrow window, the very last to be analyzed in detail, the Higgs appeared. Once again, the late-night inventions of quantum physicists turned out to be reality.

But as important as the discovery of the Higgs was for the foundation of our understanding of the physical universe, it actually raised far more questions than it answered.

THE EDGE OF KNOWLEDGE

In the first place, we don't have the slightest understanding of why the Higgs boson, and the background field it sets up throughout space, have the properties they have. It seems completely ad hoc—a convenient addition to the theory that makes it work and allows the physical world to exist with the properties it does, allowing stars, galaxies, people, and essentially all forms of matter to exist—without any underlying compelling mathematical rationale.

Moreover, everything we know about quantum physics suggests the observed scale of the Higgs, including its mass and the energy scale of electroweak unification, actually has a value in a range where it probably *shouldn't* exist.

The point is that the Higgs is the first example of a fundamentally spinless particle. Such particles are particularly sensitive to the effects of quantum processes that occur on small scales, spawning what are called virtual particles.

The laws of relativity combined with quantum mechanics imply that on small scales, virtual particles can spontaneously appear and quickly disappear invisibly and with impunity. While we cannot measure these particles directly, their brief appearance and disappearance affects the properties of particles we can measure in a way we can precisely calculate. And these effects have now been measured to over ten decimal places, with theory agreeing with observation.

The problem is that when it comes to the Higgs particle, the effects of virtual particles is such as to drive up what should be its measured mass. In principle, its mass would be driven up to infinitely high values, but instead, if the theory gets embedded in some more fundamental theory at some high energy scale, the Higgs mass would stabilize at that scale.

The most obvious high energy scale where we think some new physics must intervene is the scale where quantum mechanical effects in gravity should become important, the so-called Planck scale. As previously mentioned, this is some nineteen orders of magnitude in energy higher than the mass of the proton. But the Higgs mass is

88

seventeen orders of magnitude smaller than this scale. Something—
we don't know what—must enter into physics to solve this problem.

This problem has become known as the hierarchy problem in
physics because of the huge hierarchy of scales between the electroweak unification scale and the scale of quantum gravity.

The hierarchy problem has motivated a great deal of theoretical
work in an attempt to come up with some new physics that might
stabilize the electroweak scale. Probably the most elegant attempt
has involved proposing a fascinating new symmetry in nature, called
supersymmetry.

As I have described, elementary particles can behave like they
are spinning because they carry innate angular momentum. Spin ½
particles like electrons and protons are called fermions, after Enrico
Fermi, who described the quantum statistics of these particles. Other
particles, like photons and the W and Z, have twice the spin angular
momentum, or spin 1. These particles are called bosons, after the
Indian physicist, Satyendra Nath Bose, who along with Albert Einstein, described the quantum statistics of these particles.

Bosons and fermions are quite different, and indeed behave oppo-
sitely in many respects. The Pauli exclusion principle in quantum
mechanics says that no two identical fermions can occupy the same
quantum state at the same time. Bosons, on the other hand, like to
be in the same state, and there is a phenomenon called Bose-Einstein
condensate that describes this. This means even a macroscopic collection of bosons can be in the same quantum state at one time, behaving
quantum mechanically instead of classically. When it was first
observed in the laboratory, it was so remarkable the experimentalists
who observed it won the Nobel Prize. The background Higgs field
that permeates space is actually a coherent cosmological Bose-Einstein condensate of Higgs particles, so in that sense, Bose-Einstein
condensate is responsible for our very existence.

As different as bosons and fermions are, a novel mathematical symmetry, supersymmetry, was developed to connect them. If

supersymmetry were a manifest symmetry of nature, for every type of bosonic particle in nature there would exist a corresponding fermionic particle, with the same mass, charge, and so forth.

Of course, the world looks nothing like this, so one may wonder why we even talk about this symmetry. The reason we do, besides its mathematical elegance, arises from the impact of supersymmetry on the effects due to virtual particles. Virtual bosons and fermions provide quantum contributions of opposite signs to things like the mass of the Higgs particle. If supersymmetry were manifest, the virtual quantum contributions of bosons and fermions to the Higgs mass would cancel out exactly, so that no virtual quantum processes would impact the Higgs mass, causing it to become very large, approaching the Planck scale. It would solve the hierarchy problem.

Can we have our cake and eat it too? Can supersymmetry solve the hierarchy problem even if it is not manifest in the world we observe? Yes. The background Higgs field causes the particles we observe to have different masses, causing the world we observe to be very different than the description of the world on fundamental scales, where all particles would be essentially massless. Similarly, it is possible to imagine a related phenomenon involving another kind of weird background condensate that could cause the supersymmetric partners of all the particles we observe to get masses that are so large that we haven't yet discovered them in accelerators.

Not only would this explain why we don't observe supersymmetry as a symmetry of nature on our scale, it could also explain why the electroweak scale is what it is. If the mass difference between particles and their supersymmetric partners was of the same order as the electroweak mass scale associated with the W, Z, and Higgs, then virtual quantum effects between bosons and fermions wouldn't cancel out completely. The amount by which they wouldn't cancel out would be in the order of the mass difference between particles and their superpartners. Thus, quantum effects would give a contribution of the order of the observed Higgs mass to the Higgs particles.

Supersymmetry seemed like such a compelling idea, both as a fundamental symmetry of nature and because of its potential resolution of the hierarchy problem, that the Large Hadron Collider (which discovered the Higgs), was actually expected to possibly discover supersymmetric partners of ordinary matter before it discovered the Higgs. Alas, it didn't.

Absence of evidence is not always evidence of absence, however, and supersymmetric extensions of the Standard Model have enough flexibility that they could have avoided detection so far. But the longer we continue to probe, and the longer we don't discover any super-partners of ordinary matter, the harder it will be for supersymmetry to remain a viable solution to the hierarchy problem. We thus find ourselves in the situation of knowing there is a problem, but not yet knowing if we have the correct solution.

Incidentally, since supersymmetry is an integral part of string theory, there is even more riding on its possible existence. A possible theory of quantum gravity may also hang in the balance....

* * *

I have discussed electrons, protons, and neutrons as making up atoms, but of the three, only electrons, as far as we know, are truly fundamental. Protons and neutrons are comprised of more fundamental, fractionally charged particles called quarks.

Quarks were first proposed as mathematical constituents of strongly interacting particles like protons and neutrons in order to resolve various puzzles associated with the classification of the growing variety of subatomic particles being discovered at accelerators in the 1950s and '60s. I use the phrase "mathematical constituents" to reflect the fact that the quark model was viewed as an abstraction embodying mathematical symmetries that probably resulted from some deeper unknown underlying physics; in other words, quarks themselves were not necessarily viewed as real particles. This view was true for one of the proposers of the model, including physicist

Murray Gell-Mann, who gave them the fanciful name based on a quote from James Joyce's *Finnegans Wake*.

One of the reasons that quarks were viewed with suspicion is that no fractionally charged particle had ever been observed. The problem here was deeper than the problem associated with the earlier non-observation of atoms. For atoms, there were no tools at the time to probe matter on such small scales. For quarks, however, the situation was different. Particle accelerators were colliding energetic proton beams with matter, creating a host of new strongly interacting particles in the process, but nothing resembling a fractionally charged quark emerged from the collisions.

The reality of quarks emerged in 1969, and as you may have guessed, neutrinos played a role. Experimenters at the Stanford Linear Accelerator Laboratory in Palo Alto used a beam of electrons to probe the detailed nature of protons. Electrons were used because, like neutrinos, they do not feel the strong force that governs the interactions of protons and neutrons. Thus, their interactions with protons would be solely due to electromagnetism, and since the strong force was not fully understood at the time, the results of experiments in which electrons scattered off protons could be more easily analyzed.

Recall that when Rutherford scattered alpha particles off atoms, he was surprised to find a single, heavy, incredibly dense nucleus at the center of atoms. In the case of electrons scattering off protons, the result, while different, was equally surprising. The protons seemed to be made up of smaller particles that appeared to be moving about freely inside the photon, like particles in a box.

At first, it was not clear that these were the quarks envisioned by Gell-Mann, and Richard Feynman dubbed them partons. In order to determine their properties, other experimentalists bombarded protons with high energy neutrinos, which only interact via the weak interaction. By combining the results from the electron scattering experiments and the neutrino scattering experiments, it was confirmed that indeed the partons had fractional charges—exactly as predicted in the quark models. Quarks were real!

There were two big problems, however. First, protons and neutrons interact via the strongest force in nature, the so-called strong force. But the quark-like objects inside protons probed by the electron-scattering experiments appeared to be moving more or less freely, with minimal interactions. The second problem was even more dramatic. If these particles existed inside of protons, why hadn't a single scattering experiment knocked one out, revealing the existence of a bare quark?

The resolution of these conundrums was as magnificent as it was unexpected. Theorists had built a theory based on the quark model in which the interactions of quarks mimicked the electromagnetic interactions of charged particles, but in which there were three different kinds of charges. For lack of better terminology, physicists called them color charges, based on the analogous fact that there are three primary colors in art.

However, since the strong interactions of quarks are much stronger than electromagnetism, physicists didn't have adequate mathematical tools to fully analyze the quantitative implications of this theory.

In 1972, David Gross, Frank Wilczek, and independently, David Politzer derived an amazing theoretical result. While analyzing how the strong interactions between quarks would depend on the distance between the quarks, they discovered that the interaction strength would get *weaker* as the quarks approached each other. This would mean that on the high energy scales being probed in the electron scattering experiments, which probed the interior of the proton on very small scales, quark interactions would be suppressed. This would explain the surprising electron- and neutrino-scattering experiment results, where the "partons" inside the proton appeared to be almost non-interacting. Gross and Wilczek dubbed this remarkable property of the strong interaction "asymptotic freedom."

There was another side to the coin of asymptotic freedom. If the interaction strength between quarks becomes weaker at small distances, then it should grow stronger at larger distances. This suggests a possible new property of the strong interaction, called confinement, that could explain why no free quark has ever been observed. If the interaction strength continues to grow with distance, like a strong rubber band, then quarks could forever be bound together.

Numerical experiments on the strong interaction in a range where the interaction strength is too strong for traditional methods used by physicists to treat elementary particle interactions suggest the theory is confining. There exists no absolute mathematical proof of this fact, however. The Clay Mathematics Institute has offered a prize of $1 million to anyone who can provide such proof. For the moment, confinement remains a known unknown.

Asymptotic freedom allowed experimentalists to compare theoretical predictions with observations, because at short distances, when high energy probes can explore the nature of protons and neutrons on small scales, the strong interaction becomes weak enough that theorists can make precise mathematical predictions. The result was that the theory, called quantum chromodynamics in analogy to the quantum theory of electromagnetism called quantum electrodynamics, became firmly established as a successful theory of the strong interaction.

With the advent of quantum chromodynamics, and the successful electroweak unification of the weak and electromagnetic interactions, physicists obtained a working knowledge of three of the four known forces in nature, with only a quantum theory of gravity outstanding. While experimental verification took decades, the Standard Model as a theory had achieved its present form by 1972.

* * *

In 1929, the British theoretical physicist Paul Dirac figured out a way to unify quantum mechanics and special relativity, in turn producing a quantum theory of electromagnetism called quantum electrodynamics. It was a remarkable achievement that placed Dirac in the ranks of the greatest theoretical physicists of the twentieth century. It also made a prediction that he found embarrassing.

The equation he derived, now called the Dirac equation, had an unexpected implication. In modeling the relativistic quantum states of electrons, the equations required a new set of states—of seemingly negative energy. These states could alternatively be considered positive energy states, but only by assuming the existence of particles with an electric charge equal and opposite to that of the electron, but with otherwise equal mass and other properties.

Dirac was flummoxed. He proposed that perhaps seemingly empty space was full of a sea of negative energy electrons, now called the Dirac sea, and that they were undetectable. With enough energy, one could kick an electron out of this sea and produce a real observable electron. That would leave one less electron in the sea, which would therefore appear as if it left an equal and opposite positive "hole" (the absence of a negative electron) in the sea. He argued that perhaps the interactions of that hole with other electrons in the sea would cause it to behave like a heavy positive particle, which he identified with the proton. But it was quickly shown by Pauli and others that this argument didn't fly.

Dirac's confusion didn't last long. Within a year of the development of his theory, the American physicist Carl Anderson, examining the tracks of cosmic ray particles in detectors, discovered among the cosmic rays a track corresponding to a particle that resembled an electron in every way except that it was positively charged. It was quickly recognized that this particle, dubbed a positron, was precisely the particle predicted by Dirac's theory. Dirac himself later said that his equation was smarter than he was.

It is now understood that as a consequence of the combination of quantum mechanics and special relativity, all elementary particles like electrons and quarks must have what are now called antiparticles—partners with equal mass and opposite charge. Some neutral particles, like photons, the carriers of electromagnetic interactions, can be their own antiparticles. Just as positrons must exist, so must antiquarks, which can make up antiprotons, and so on. All such antiparticles have now been detected, and indeed it has become routine to produce them in elementary particle accelerators. In fact, physicists can create beams of antiparticles and collide them with particles in order to further explore the properties of matter and radiation.

Antiparticles seem like the stuff of science fiction, but they are not. When a particle and antiparticle collide, they can convert their mass-energy into pure radiation, but other than that, antiparticles behave essentially identically to their particle partners. They fall down in a gravitational field just like particles. Antihydrogen, comprised of bound states of a positron and an antiproton, has essentially exactly the same atomic spectrum as hydrogen. We only call antimatter by that name because we happen to be surrounded by matter rather than antimatter. As I have said in my book *The Physics of Star Trek*, if the earth was made of antimatter, then anti-lovers might sit in anti-cars under an anti-moon making anti-love. I also said there that the only reason antimatter seems strange and exotic it is because we don't see it often. Belgians too may seem strange and exotic. They aren't, but they may seem that way because you rarely meet someone from Belgium. Once I gave a talk in Belgium and found out they didn't appreciate that joke.

This of course then leads to a deeper question. Why do we live on an earth and not an anti-earth? More specifically, how come when we look out at our neighboring planets, stars, and galaxies, we only see evidence for matter and not antimatter? To be sure, we detect some antiparticles amidst the flotsam and jetsam of high energy cosmic rays impinging on earth, but they are rare and exotic. How come we appear to live in a universe of matter?

You may not wake up every morning wondering why this is the case, but once physicists began to think of applying the microscopic laws of physics to our universe, this question begged for an answer.

First, a bit of nomenclature. Protons and neutrons are examples of what are called baryons. The asymmetry between matter and anti-matter, when it comes to the matter made from protons and neutrons that makes up everything we see, is called the baryon asymmetry of the universe.

The problem is that any sensible picture of the early universe should have had equal numbers of particles and antiparticles. This is because, as we have seen, the very early universe was hot and dense. When the temperature was far higher than the energy associated with the resting mass of particles, radiation could be converted into matter and vice versa. But when neutral radiation converts to matter, equal numbers of particles and antiparticles are created, so that the total charge created is zero. This means that at very high temperatures, there should have been an equal number of particles and antiparticles in the dense gas of particles and radiation.

If there were an equal number of baryons and antibaryons at early times, however, I wouldn't be here to write this, and you wouldn't be around to read it. This is because the strong interaction between baryons and antibaryons is strong enough that all baryons and anti-baryons would have been able to annihilate each other by the present time, leaving a universe composed of nothing but radiation.

Working backwards from the present time, we can see that this was very nearly the case. In the universe today, there are roughly between one and ten billion photons in the CMBR for every proton in the universe. This means that all one would have needed to produce this observed ratio today would have been a slight excess of one extra baryon for every billion or so baryons and antibaryons at very early times. The billion baryons would have been annihilated with the billion antibaryons, producing photons that now make up the CMBR. With the one extra unpaired baryon remaining in each region, these baryons were enough to produce all the stars and galaxies we see in the universe today.

Either the universe was somehow created with a very, very small excess of matter over antimatter, which seems hard to understand, or somehow dynamical processes since the beginning of time created such a small excess. The challenge for modern cosmology is to figure out how that might be possible.

In 1967—before the bulk of the physics community began to seriously contemplate using elementary particle physics to understand the early universe—the Soviet physicist Andrei Sakharov, one of the fathers of the Soviet hydrogen bomb and who later became famous as a Russian dissident, wrote a prescient physics article outlining exactly what was required to generate baryon asymmetry in the early universe. It wasn't pretty, because none of the ingredients existed in the known laws of physics at the time. They were:

1. *Interactions that can create or destroy the net number of baryons.* These don't exist in everyday physics. If they did, then protons would be unstable; they could decay into light particles like positrons and neutrinos or photons.

2. *A departure from thermal equilibrium in the early universe.* In thermal equilibrium, the rate of reactions that go in one direction, like the creation of baryons, say, must be exactly equal to the rate of reactions that go in the other direction, like those that would destroy baryons. So even if baryon-violating interactions existed, if the universe were in thermal equilibrium, then no total baryon asymmetry would be created if there was none to begin with.

3. The third ingredient is more subtle, but basically it is *a violation of the symmetry between particles and antiparticles.* I indicated earlier that aside from their opposite charges, antiparticles have essentially the same interactions as their particle partners. If this were exactly true, then for every process in the early universe in which a particle interacted with other particles to either create or destroy a baryon, precisely the opposite would

happen for processes involving antiparticles. Once again, if there were no baryon-antibaryon asymmetry to begin with, none could develop as long as this symmetry holds.

The symmetry between particles and antiparticles is a bit subtle and worth exploring in a little more detail. We really need to consider a combination of two separate symmetries that can exist in nature. In order for nature to appear the same when particles and antiparticles are interchanged, first one could change all positive charges to negative charges and then vice versa. (Symmetry under this interchange is called charge conjugation symmetry and is labeled C). Next one could perform an interchange of left and right, which we have already encountered, and which is called parity symmetry, labeled P. The reason both operations (CP) need to be performed can be understood by considering a simple example. Say we have an electron and a positron moving in opposite directions, with the electron moving to the left and the positron to the right. Performing a C operation would result in a positron moving to the left and an electron to the right. Then performing an interchange of left and right, P, would result once again with an electron moving to the left and a positron moving to the right.

It is also a remarkable fact that the combination of relativity and quantum mechanics tells us that nature must be invariant under simultaneous reversal of charge (C), parity (P), and direction of time (T). So if CP is violated, as would be required to generate a baryon-antibaryon asymmetry in the universe, then this implies that T is violated as well, so that the combined operation of C, P, and T would leave physical systems unchanged. So, evidence for CP violation is the same as saying, remarkably, that the laws of physics distinguish between forward and backward directions of time at some fundamental level.

Of the processes required by Sakharov's theory, the first to be shown to exist was perhaps the strangest. CP violation was discovered in 1964, to the great surprise of the physics community, in the rare decays of exotic particles called kaons. These particles contain a quark that is different than the quarks that make up protons and neutrons, called the strange quark by Murray Gell-Mann. The discovery that CP was violated by the weak interaction in the mixing of strange particles and antiparticles shocked the physics community once again, having been rocked only eight years earlier by the discovery of parity violation.

CP violation was a great surprise because it is so rare and hence so small. It took almost thirty years before any other examples of CP violation were discovered in even more exotic particle systems.

The possible origin of such a small effect remained a mystery for almost a decade after the experimental discovery of CP violation. Then, in 1973, Makoto Kobayashi and Toshihide Maskawa made a bold proposal, which followed an earlier and equally bold proposal. Around the same time as the discovery of CP violation, the theoretical physicists Sheldon Glashow and James Bjorken speculated that a fourth quark, which they called the charm quark, might exist. Glashow and collaborators, in 1970, provided further mathematical arguments for why it might exist, and in 1974, the first particle containing a charm quark was discovered.

The proposal of a fourth quark, the charm quark, was suggested so that quarks could fall into two "families," just like the two families of particles, leptons (consisting of the electron and its heavier cousin, the "who ordered that" muon) and their associated neutrinos. In this way, weak interaction, which can convert electrons and muons into their respective neutrino partners, could be pictured to similarly convert the up and down quarks (which make up protons and neutrons) into each other, and the strange and charm quarks in the second family into each other.

Kobayashi and Maskawa built on Glashow and his collaborators' proposal of a charmed quark by proposing that there should exist a third family of quarks, now known as the bottom and top quark. Their proposal, purely theoretical, was based on a simple mathematical result. If there were three families of quarks, it was possible to show that the quarks' weak interactions could cause the quarks to mix each other in a way that would allow a CP-violating term to naturally result in these interactions. With two families one could show that no such term could independently appear.

Imagine the chutzpah: predicting a whole new family of quarks, motivated by the esoteric mathematical properties of an object called a matrix, which would allow the existence of a very small CP-violating parameter in the weak interaction.

Once again, nature played along. In in 1975, the tau lepton, a third heavy cousin of the electron, was discovered, along with its own neutrino. Then, to complete the symmetry between quarks and leptons, in 1977, the bottom quark was discovered and in 1995, the top quark was discovered.

Now there were three families of elementary particles in nature, though only the first one produced any (and in fact, all) of the objects with which we are familiar. "Who ordered that" squared?

With three families of quarks and leptons, there are twelve possible masses for these twelve particles. There are three different ways that the weak interactions could couple combinations of quarks and three different ways the weak interaction could couple combinations of neutrinos. What is the origin for the values of all these different parameters, each of which has been measured with varying levels of precision? We don't know.

Does it stop at three? Could there be more, yet heavier families of quarks and leptons waiting to be discovered? We actually strongly suspect there aren't because of the mass of the Higgs particle, which is comparable to the mass of the heaviest quark, the top quark. If the masses of particles arise because of their coupling to the Higgs field, it is hard to implement that scheme if their mass is much larger than that of the Higgs.

This argument is suggestive, but since we don't really understand what, if anything, determines the nature of those couplings for the different families, who knows?

This somewhat lengthy digression into CP violation and the existence of three families of quarks and leptons is relevant, not just because it illustrates the heart of our knowledge, and lack thereof, about the particles making up all matter we know about, but because it is also relevant to our understanding, and our lack thereof, of the processes producing the observed baryon asymmetry of the universe.

The discovery of CP violation in the Standard Model of particle physics may seem like it provides one key ingredient that might explain the baryon asymmetry, but alas, it presents an obstacle to any explanation. The observed CP violation in the weak interactions of quarks is too small to explain how a baryon-antibaryon asymmetry even as small as one part in a billion could have been generated at early times.

Nevertheless, rather remarkably, it turns out that the other two ingredients of Sakharov's recipe for baryogenesis also exist in the Standard Model. While the weak and electromagnetic interactions were unified at small scales and very high temperatures as the universe cooled, at some point they began to diverge in strength. This means that there was a phase transition in the early universe, from a phase where electromagnetic and weak interactions were essentially identical, to one in which they differed. Whenever there is a phase transition in nature, there is the possibility of a departure from thermal equilibrium, as in the example I described earlier, when water cools below zero degrees Celsius on a busy street in the winter and then suddenly freezes.

Moreover, exotic processes first uncovered by the brilliant Dutch theoretical physicist Gerard 't Hooft allow for the violation of baryon conservation even in the Standard Model. Happily, at least for our continued existence—built as we are from protons and neutrons—the effect is unbelievably small at low temperatures. However, in the early universe, such effects could have occurred with wild abandon. That sounds like a godsend, but it isn't. If these processes are in thermal equilibrium at high temperature, their net effect would generally be to erase any preexisting baryon asymmetry.

There is one final problem with the Standard Model, which is also related to considerations of CP violation. The small CP violation observed in the weak interaction feeds into the strong interaction, which can result in arbitrarily large CP-violating effects in protons and neutrons, none of which are observed. This problem, called the strong CP problem, has been with us since the 1970s, and we still don't know the answer, although it is clear that it requires new physics beyond the Standard Model.

Generating the matter-antimatter asymmetry in the universe, dealing with CP violation, and understanding the makeup of fundamental particles all point to the need for an understanding of physics beyond the Standard Model.

Beyond this, in the absence of these outstanding puzzles, there is one feature of the observed nature of the universe that cries out for explanation. There are currently four known forces in nature: gravity, electromagnetism, the weak force, and the strong force. They vary in strength by over forty orders of magnitude, with gravity being the weakest force, and the strong force, unsurprisingly, the strongest. Are there other as-of-yet undiscovered forces?

The weak and electromagnetic forces have been unified together through the discovery of the W, Z, and Higgs particles, and this begs the question that helped motivate Alan Guth's first proposal of inflation: is it possible that all the known (and any unknown) forces in nature are unified at some high energy/small distance scale?

In 1974, after the discovery of asymptotic freedom and the achievement of the basic theoretical framework of electroweak unification, and even before the discovery of the W, Z, and Higgs particles, bottom quark, top quark, and tau lepton, two key developments strongly suggested to physicists that at least the three non-gravitational forces in nature might be unified at a high energy scale.

Once it was recognized that the strong force gets weaker as one uses higher energy probes to explore it on smaller length scales, a similar examination was applied to electromagnetism and the weak force. It had been known for decades, if not often remarked upon, since the work of the great Russian theoretical physicist Lev Landau, that the strength of electromagnetism is also scale-dependent, and that electromagnetism increases in strength on smaller scales. Above the electroweak scale, about one hundred times the mass of the proton, corresponding to distances about one hundred times smaller than the size of the proton, the strength of the two couplings associated with the combined electromagnetic and weak forces could be calculated, and a striking pattern began to suggest itself. With the weak and electromagnetic forces getting stronger, and the strong force getting weaker, could it be that all three forces would unify in strength at some ultra-high energy scale?

Early calculations by Steven Weinberg and Howard Georgi, among others, strongly suggested such a possibility with a unification scale about fifteen orders of magnitude in energy greater than the rest mass of the proton, or about fifteen orders of magnitude in size smaller than the proton. As fascinating as this possibility was, this miniscule scale was well beyond anything accelerators then, or now, could probe directly.

At the same time, Sheldon Glashow, who along with Weinberg had helped spearhead the unification of the weak and electromagnetic forces, recognized another fascinating mathematical property of the known forces and particles.

As I have mentioned, each force in nature is associated with a certain type of symmetry, called a gauge symmetry. Mathematically, one can associate a quantity called a lie group with every such symmetry. The bigger the group structure, the more particles are involved in conveying the force. For electromagnetism, there is one particle: the photon. For the weak force there are two: the W and Z. The strong force has eight particles, called gluons, communicating interactions between quarks.

Glashow, along with collaborator Howard Georgi, noted that all of these symmetry groups could be combined in a simple elegant structure that would not only accommodate all the known forces in nature, but could accommodate all the known elementary particles as well.

Moreover, if this were realized in nature, there would be new interactions at this super high energy scale that would allow protons to decay into light particles like electrons and neutrinos. Because the interactions involved new physics at very high energy scales, the effect of these interactions on the scales we currently measure would be infinitesimally small. Protons would indeed decay, but with a lifetime in excess of 10^{30} years, which is twenty orders of magnitude longer than the current age of the universe. When the temperature of the universe was comparable to the scale of these new interactions, however (when the universe was about 10^{-35} seconds old), these baryon-violating interactions could have occurred with wild abandon. And, if the strong, weak, and electromagnetic interactions unified at this scale, then when the temperature fell below this scale there could have been a phase transition that would have resulted in out-of-equilibrium effects. Finally, with all the new particles and interactions, new sources of CP violation would have been easily possible. The long sought-after fundamental possible sources of the baryon asymmetry of the universe seemed at hand.

The smell of grand synthesis was in the air, and it was therefore only appropriate that these new theoretical proposals were given a new name in 1978: grand unification.

I remember the excitement associated with grand unification as the particle physics community began to focus on these exciting possibilities around the time I began graduate school. I even remember attending the first workshop on grand unification. My friend Joe Lykken (now a deputy director of Fermilab) and I arrived early at the conference site and we were asked by the conference staff what grand unification was. We couldn't resist telling them it was a new religion, the Grand Unification Church, with High Priests Sheldon Glashow and Steven Weinberg.

Unfortunately, like an impressionist painting, while the grand unification picture was sublime when seen from afar, as one began to examine it more closely, the picture became blurry and cracks began to emerge.

The first crack had to do with proton decay. Motivated by these exciting developments, experimenters built huge underground tanks in working mines, filled them with ultra-purified water, and surrounded them with photo-tubes sensitive to any light that might emanate from inside the tanks. The reasoning was that if an average proton lived 10^{30} years, then if one assembled 10^{30} protons in a single tank one would expect to see, on average, one proton decaying each year. The scattering of the decay products would then produce significant radiation as they traversed the tank.

The tanks were built and installed in Japan, the US, and elsewhere, and experimenters waited for a signal. And waited. And waited. It has now been over forty years since the first proton decay detectors came online, and so far, no signal. This does not kill the idea of grand unification. The proton lifetime depends on the details of models and the scale of unification. While the original Glashow-Georgi model hasn't survived, there are many other possibilities, for, if the unification scale increases by a factor of even two to four, the proton lifetime increases by an order of magnitude or more.

A more serious problem, at least for the simplest idea of grand unification, is that once more detailed measurements of the strength of the three non-gravitational couplings were performed at high energies and more detailed theoretical calculations were performed, it became clear that, at least in the context of the known physics of the Standard Model, their interaction strengths did not converge together at a single high energy, as would be expected if they were unified together.

Fortunately, a possible solution for this problem jumped up at about the same time the problem itself was discovered. Recall that to solve the hierarchy problem—the vast disparity between the scale of electroweak unification and the scale at which quantum effects become important in gravity—physicists had proposed the existence of a new symmetry of nature, supersymmetry. A host of new supersymmetric particle partners of ordinary matter could exist near the electroweak scale, ensuring its stability. The existence of these new particles would change the calculations of the convergence of the interaction strength of the three non-gravitational forces, and when this possibility was incorporated into new calculations, another miraculous result emerged. The three interaction strengths would now nicely converge at a single scale, now about sixteen orders of magnitude larger than the rest mass of the proton. This increased scale would also explain why no proton decay results had been observed.

Unfortunately, however, as I have described, to date none of the super-partners of ordinary matter have been observed at the LHC. That doesn't kill the idea, of course, because there remains flexibility, but it does at the very least dampen hopes, along with reducing the phase space of allowed models. We will see if supersymmetry and grand unification rise again or are relegated to the dustbin of history. Right now, the evidence for both is so strongly suggestive that most physicists are betting on a discovery, perhaps in the new runs of the LHC, which has just turned on again as I am writing this.

The development of grand unification was the intellectual father of the effort to unify gravity along with the other forces of nature, with the rise of string theory and all that that has entailed. But we shall see. Right now, we don't know what new physics will emerge to resolve these outstanding mysteries.

There is one important feature of grand unification and the baryon asymmetry of the universe that is relevant to a question I discussed at length earlier: the far future of the universe. If protons do decay, albeit with a lifetime of 10^{32} years, then as Sheldon Glashow once stressed,

"Diamonds are not forever." Ultimately, matter itself will be unstable. If we wait long enough, matter in the universe will disappear. There will be no protons or neutrons, no atoms, no planets, no stars. Could there be any interesting physical processes in such a cold, dark universe, perhaps interesting enough to allow some exotic life forms to merge and exist? We don't know.

* * *

Earlier I stated that neutrinos have played a role in essentially every new development in our understanding of matter and the forces of nature on a fundamental scale, but you will note that they haven't appeared to play a role in this recent parable about the ongoing mysteries of fundamental physics, including the baryon asymmetry of the universe, the number of families of particles, CP violation, and grand unification theory. Now it is their turn to enter the story.

All the evidence for physics beyond the Standard Model—the hierarchy problem, the baryon asymmetry of the universe, and grand unification—has been indirect so far. To date, the only direct evidence that the Standard Model is incomplete has come from a measurement of neutrinos.

In the Standard Model, neutrinos are essentially massless. They are only left-handed, meaning their spin angular momentum points in the opposite direction of their motion. But for a massive particle, which always moves slower than the speed of light, I can always overtake the particle and move past it if I am moving fast enough. But if I do, then in my frame the neutrino will be moving in the other direction, though the direction of its spin angular momentum will not have changed. A left-handed particle will have become a right-handed particle. Massive particles must therefore exist in both left- and right-handed states.

In the 1960s and '70s, one of the most frustrating outstanding puzzles in particle physics and astrophysics was called the solar neutrino problem. Detailed observations of the sun, combined with well-understood nuclear physics developed in the 1940s, had previously solved one of the biggest early twentieth century puzzles in science: why does the sun shine? The discovery of nuclear fusion had not just led to the development of new weapons of mass destruction here on earth, it had allowed physicists to discover and calculate the processes that power the sun and will continue to power it for at least five billion more years.

The conversion of protons—the nuclei of hydrogen—into nuclei of helium releases about twenty million times more energy than the most energetic chemical reactions. We know the power output of the sun, and we can work backwards to calculate the rate of nuclear reactions in its core. We can't probe those reactions directly with our telescopes, however, because they cannot peer inside the sun. As I have described, the photons that emanate from the solar surface have taken almost a million years to escape the sun, randomly walking and colliding with atoms in the solar interior, and thus they too cannot provide any direct probe. Fortunately, however, there is such a probe in the form of neutrinos.

Neutrinos are emitted in a number of the nuclear reactions powering the solar interior, and because neutrinos interact so weakly, they escape the sun in a matter of seconds, unscathed. Every second of every day over ten thousand billion neutrinos coming from the sun impinge on your body and go through it without knowing you were there.

In 1965, a brave (or foolhardy, depending on how you view it) chemist named Ray Davis took the early estimates of the neutrino flux from the sun seriously enough to see if he could build a detector to detect them. In a mine in South Dakota, almost a mile underground, he built a one hundred thousand–gallon detector full of cleaning fluid. Of the billions and billions of neutrinos that traversed his detector, calculations suggested about one neutrino a day would collide with the nucleus of a chlorine atom and turn it into the nucleus of an argon atom. And amazingly, Davis decided he could detect thirty atoms of argon produced per month amidst the one hundred thousand gallons of perchloroethylene.

To the amazement of the physics world, the experiment worked. But there was a problem. Over the next twenty years, Davis continued to detect only about 30 percent as many neutrinos as the standard solar model predicted were being produced.

It was a very hard experiment, and many physicists presumed that perhaps the experiment was simply not able to detect neutrinos with the accuracy that was claimed. Other physicists, distrustful as they are of astrophysics, assumed the solar model calculations were probably wrong. After all, the interior of the sun must be hellishly, both literally and figuratively, complex.

I spent a number of years with my students and colleagues trying to figure out if errors in the solar model could explain the Davis results while remaining consistent with other astrophysical data, but that proved difficult. Other groups also confirmed this result.

Perhaps then, the problem was not with the sun, or Davis's detector, but with neutrinos. As long as neutrinos were massless, the Davis detector was guaranteed to detect the predicted neutrino flux. But, if neutrinos had a mass, even a very small mass—especially a very small mass—then a phenomenon that had been observed in other elementary particle systems could occur for neutrinos: oscillations between different neutrino types.

If even just two different types of neutrinos had different masses, then it was possible that electron neutrinos, which are the neutrinos emitted in nuclear reactions, could convert, on their way from the sun to earth, into their cousins, muon neutrinos. Since the Davis detector could only detect electron neutrinos via the nuclear reactions they induced in chlorine, then the detector might therefore naturally detect a paucity of neutrinos.

I am not sure how many of us really believed this was the case, but thousands of physics papers, including at least half a dozen by me and my collaborators, explored this possibility. My colleague Sheldon Glashow and I even named one type of oscillation—in which the neutrino masses would be "just so" to allow just one oscillation between the sun and earth—as "just-so oscillations," after Rudyard Kipling's *Just So Stories.*

The way to prove the possibility of neutrino oscillations was to build a neutrino detector that would detect all neutrino types and not just one type. Such a detector, using heavy water and called the Sudbury Neutrino Observatory (SNO), was built in a mine in Sudbury, Canada. Sure enough, in 2001, SNO reported its results: the flux of all neutrino types matched the solar model predictions. Electron neutrinos had oscillated. Neutrinos had a mass. Further results from using atmospheric neutrinos created in the collisions of cosmic rays with our atmosphere demonstrated that not only did electron neutrinos oscillate into muon neutrinos, but muon neutrinos also oscillated into tau neutrinos. All neutrino types appeared to have a mass.

These were truly profound results, not just for their significance for astrophysics, but also because this remains the only direct physics result that cannot be accommodated within the framework of the Standard Model. For neutrinos to have a mass means that right-handed neutrinos must exist. This suggests one of two possibilities: either that the neutrino is its own antiparticle so that left-handed neutrinos and right-handed antineutrinos are just different parity states of the same particle, or there are new exotic right-handed neutrino states which, according to the Standard Model, don't interact at all with any particles of normal matter. Both possibilities imply dramatic new physics.

Some of you might be reminded of dark matter when you read about new particles that don't interact with normal matter, and for a long time, neutrinos seemed like the prime candidate for the dark matter dominating the mass of galaxies. But unfortunately, existing constraints on their masses rule out this possibility.

Aside from that, however, the existence of neutrino masses has significant implications for both grand unification and mechanisms to produce the observed baryon asymmetry of the universe.

Physicists had wondered about neutrino masses for some time, and the big question that arose was: if neutrinos are indeed massive, why are their masses are so small? Even in the 1980s, constraints on neutrino masses were such that they had to be, at most, five hundred thousand times smaller than the mass of the next lightest particle, the electron.

The answer was a mechanism called the seesaw mechanism. If new right-handed neutrino states existed with a very large mass, corresponding to the grand unified scale, then their left-handed counterparts could naturally have extremely small masses, just in the range to produce the kind of neutrino oscillations that might have explained the solar neutrino problem.

So, perhaps the strongest direct evidence that there is new physics at grand unified scales comes from neutrinos. But there is an added benefit to this separation of right-handed and left-handed neutrinos into separate mass scales. It implies that neutrinos must be their own antiparticles, also known as Majorana particles. But since neutrinos are leptons, and antineutrinos are antileptons, identifying leptons and antileptons means that neutrino masses would violate the physical symmetries that normally distinguish particles from antiparticles.

This could solve the problem of the matter-antimatter asymmetry as follows: while the physics of the weak interaction at high temperatures could wipe out any initially generated baryon-antibaryon asymmetry in the early universe, it cannot erase a lepton-antilepton asymmetry that might have existed. If such an asymmetry is generated early on, then this asymmetry could then be converted into a subsequent baryon asymmetry through exotic new interactions of neutrinos. For many physicists, these leptogenesis scenarios now represent the most realistic possibility for explaining why we ultimately might live in a universe of matter rather than antimatter.

There is one other reason why leptogenesis is attractive. If three generations of neutrinos all have a mass, then it is possible that neutrino masses can provide a new source of CP violation beyond that observed elsewhere in the Standard Model. Indeed, the CP violation in this sector could be much larger and not be inconsistent with existing experimental constraints. At least one existing experiment claims evidence for such CP violation, but the result remains controversial. Two gigantic new experiments are being built, one in the US and one in Japan, to explore this possibility.

Will neutrinos become the key that unlocks the many mysteries of the Standard Model, from the possibility of a unification of the known forces of nature to an explanation of why we live in a universe of matter? We don't know yet. Stay tuned.

* * *

Among all the mysterious features of matter at fundamental scales in nature, perhaps none is stranger than that which determines the rules that govern the behavior of matter. I am referring to quantum mechanics. It would be impossible to write about the known unknowns of nature without at least touching on the fact that the rules governing the very heart of the physical universe are crazy.

Nevertheless, just as it is impossible to even discuss biology except in the context of evolution, it is impossible to describe the modern world except in the context of quantum mechanics. It underlies everything that governs modern technology, from the computer on which I am typing to the mechanics of the smart phone I increasingly depend upon, even to the electronics that govern the functioning of my car.

Discussing all the subtleties of quantum mechanics would require an entire book. But it is possible to more succinctly outline the various key features that underlie our quantum universe:

1. Many states at once: For me, the key distinction between the quantum picture of reality and the classical picture has to do with the configurations of systems governed by quantum mechanics. Classically, if I throw a ball, it takes some well-defined trajectory determined by Newton's laws. If many

electrons are emitted by a cathode ray tube, their average trajectories will mimic those of the ball. But the specific trajectory of each electron is completely undetermined in advance. In fact, it doesn't even make sense to talk about the trajectory in advance of measuring it. That is because the electron acts as if it is taking many trajectories at once. Every measurement one can make to try and demonstrate that it actually took a specific path in advance of directly measuring its position, instead finds that no single trajectory is consistent with the data. This picture, promoted by Richard Feynman in his "path integral" formulation of quantum mechanics, most concisely captures the heart of quantum theory in my opinion.

2. The fundamental quantity in quantum mechanics is the wavefunction of an object, which, succinctly, allows an exact prediction of the probability of measuring the object in any one of the allowed states it may be measured to be in, for all times. One of the many misstatements about quantum mechanics is that it is not deterministic. This is incorrect. Quantum mechanics is based on an equation that describes the time evolution of the wavefunction. This means that if one specifies the value of a wavefunction at some initial time, its value at all subsequent times can be determined exactly, at least in principle. The wavefunction, defined more precisely, gives the probability amplitude (a complex number) of finding the system in a certain state. The square of the wavefunction gives the probability (a real number between zero and one) of measuring the system to be each of its many possible allowed states. Quantum mechanics determines these probabilities exactly. By the same token, it tells us that it is only these predicted probabilities that we can ever compare with experiments. The initial state of the system can never be determined exactly because the system can be in a combination of many states at once, a phenomenon called superposition. Such a superposition is another way

of framing the fact that a particle can follow many different trajectories as it traverses from A to B, as long as we do not measure it between those two points.

3. The order in which one measures properties of a system can determine the properties one measures. Put another way, for some properties, reversing the order in which you measure them, for example the momentum and position of a particle, will give a different result than will the original measurements. This translates into the famous Heisenberg uncertainty principle, examples of which include the fact that one cannot measure with 100 percent accuracy both the position and momentum of a quantum object, or the precise value of its energy at a given time.

4. Once a system of several otherwise distinct objects is in some fixed quantum state, coherent correlations continue to connect these objects even when they are separated, as long as the system is not disturbed. Measurements of one object can therefore instantly restrict the allowed quantum state of the other object. This phenomenon, called entanglement, is the "spooky action at a distance" that so disturbed Einstein.

5. When quantum mechanics is combined with relativity, things get even crazier, as I have already described. Quantum systems are always fluctuating, sometimes wildly. Relativistic quantum mechanics tells us that even empty space is not empty. Particle and antiparticles can spontaneously appear and disappear on timescales that quantum mechanics tells us are too short to ever be able to detect real particles. The shorter the time one considers for the evolution of any system, the more dramatic the possible fluctuations, and the higher the possible energy and mass of the virtual particles that populate what otherwise appears to be a vacuum.

These five facts capture most of the craziness of quantum theory. They imply that the sensible classical reality that most of us take for granted is an illusion. Ever since quantum mechanics was first developed to explain the behavior of atomic systems and their interaction with radiation, physicists from Einstein onward have argued that while the mathematics of quantum mechanics clearly provides correct predictions about how the world works—indeed, the most precise predictions that have ever been developed about nature—quantum mechanics is simply too crazy to be absolutely true.

Surely there must be some underlying reality in which systems behave sensibly—for example, where objects that are measured to have some property definitely had that property before the measurement—and quantum mechanics is an emergent theory that gives the right answers but hides the underlying truth. Predicting probabilities associated with measurements is just an excuse for lack of knowledge about this underlying truth. An electron cannot really be spinning in all directions at the same time. Nor is there really some probability, however small, that the photons from the light in my room circle twice around the moon before reaching my eyes. Nor is Schrödinger's cat, if it is sufficiently small, really alive and dead at the same time before I open the box to observe it.

Part of the problem in both accepting or understanding quantum mechanics has been the promulgation of misconceptions about intrinsic randomness where there is none, or about the nature of measurement, which is a subtle and complex issue and can lead to apparent nonsense when one treats the measuring system as classical but the system being studied as quantum. And alas, quantum mechanics has led to quacks and charlatans who make ridiculous claims that measurement has something to do with consciousness and that external phenomena can be influenced by consciousness in the same way that they are impacted by actual measurements, so that somehow wanting the external universe to behave some way will cause it to behave that way.

A great deal of progress has been made in trying to clarify how quantum weirdness turns into classical sense as systems become macroscopic in size and measurements are made. At the same time, all attempts to avoid the craziness of quantum mechanics—both experimental and theoretical—have failed.

In particular, it has now been convincingly demonstrated that any picture in which intermediate unmeasured systems actually behave in some classical sense, in that they must exist in some specific state rather than a possible superposition of many different states before you measure them, is ruled out.

The most convincing demonstration of this is often attributed to the late physicist John Bell, who thought deeply on these issues and proposed experimental tests based on the idea that some hidden classical theory might be consistent with quantum mechanics. A host of able experimentalists then went out and performed the (now Nobel Prize-winning) experiments, validating the predictions of quantum mechanics and apparently ruling out the alternate classical reality as a possibility.

The most compelling demonstration of this that I know of was a brilliant reworking of Bell's argument in a different context, first done by the physicists Daniel Greenberger, Michael Horne, and Abner Shimony, then refined by David Mermin, and then repackaged by one of the smartest and wittiest physicists I have ever known, my late Harvard colleague Sidney Coleman. Coleman gave a lecture entitled "Quantum Mechanics in Your Face," which you can find online. Some knowledge of physics is required to fully understand it. Martin Greiter transcribed and produced that lecture as a paper, and it is available on the online physics archive arXiv.org.

The example Coleman describes is simple and compelling. Consider some central station sending out something to three widely spaced laboratories, and the laboratories receiving that something at the same time, so there is no way the measurement in one laboratory can classically causally affect the measurement in any of the others.

Each laboratory (identified as either one, two, or three) has an identical detector, and it has two settings, A and B. If set to A, it measures one thing, which registers either +1 or -1. If set to B, it measures another thing and also registers either +1 or -1. The observers do not know what their detector is specifically measuring, nor do they know what the incoming something is. The condition described in the preceding paragraph implies that classically, whatever one observer measures, A or B, the result will not be affected by what the other observers measure.

Also, thinking classically, the observers can assume that if they measure B, then A nevertheless possesses exactly the same value that it would have had if they had chosen to measure A instead.

The observers come together after making many measurements and discover that every time one observer measures A and the other two observers measure B, the product of their measurements is always +1. *Every single time.* The individual measurements may not always be +1, but the product of the three measurements always is. Now, since B x B is always +1 whether or not B itself is +1 or −1, the experimenters conclude, by reasoning classically, that in these cases (roughly three eighths of the total, assuming the experimenters randomly and independently choose to measure A or B), A must always be +1. They next consider those occasions when they hadn't measured one A and two Bs. Since the observations are independent and cannot affect one another, they reason, thinking classically again, that had they instead measured one A, the result of the product would have had to have been +1, and hence the value of that specific A on that occasion too must have been +1. And since this is independent of whether A is measured at position one, two, or three, they conclude that if all three detectors measure property A, the answer must be +1.

Now, however, let's say that what the central station was sending out was three spin ½ particles, one to each detector. Let's also say the three particles were initially together in a specific quantum state, which was a linear superposition of the state in which all three particles had spin pointing up (in the Z direction) minus the state in which all three particles had spin pointing down. Finally, let's presume that

118

A is a measurement of the spin of the particle in the x direction, and B is a measurement of the particle in the y direction. Then quantum mechanics tells us that the product $A_1A_2A_3$ will *always* be equal to -1. (The mathematics needed to show this is relatively straightforward. See Coleman for the details.)

It is hard to find a bigger disagreement between classical reality and quantum reality. Either the product is +1 or -1, and you can guess what experiments tell us. The classical mistake the observers are making is to assume an objective reality in which each particle is in an unambiguous specific state, independent of the other faraway particles, and independent of what measurements you perform or don't perform.

Upon observing this crazy result predicted by quantum theory, experimenters who insist on thinking classically might say that there must be some faster-than-light communication between the different experiments so that somehow the results of measurements in one lab are affected by measurements in another laboratory. But this is only the case if you are thinking classically. Quantum mechanically, there doesn't have to be any superluminal communication. Once the initial state is specified, the final result is determined.

This hits at an ongoing semantic problem, which Coleman exposed, in the way many physicists and a number of philosophers try to understand quantum mechanics. People talk about the "interpretation of quantum mechanics," and some people even write books about it. But as Coleman stressed, that is getting things backwards. The world isn't classical, so any classical interpretation is a kluge, an approximation in which some weird classical behavior is imputed as an approximate substitute for the actual quantum behavior of systems. As Coleman then argued, we shouldn't be talking about the interpretation of quantum mechanics, but rather the interpretation of classical mechanics.

Quantum mechanics subsumes and replaces classical mechanics. It produces the same results as classical mechanics in the limit where quantum effects disappear, just as general relativity reduces to Newtonian gravity when gravitational fields are weak. But no one expects that the general relativistic predictions of behavior in strongly curved spaces could be sensibly described in terms of the Newtonian, flat space picture. Why do people insist on doing this for quantum mechanics? It is probably because general relativity, while it defies our direct experience, does not defy classical logic and reasoning. Quantum mechanics does, however, and that seems to be an affront that cannot be forgiven, even by physicists.

You might think the Coleman-Mermin-Greenberger-Horne-Shimony extension of Bell's original arguments would have settled the issue. The world is quantum mechanical, like it or not. But it hasn't. Respectable physicists still question whether the reality of the quantum world is just an illusion, and not just respectable physicists, but some of the world's greatest theoretical physicists too. Both Gerard 't Hooft and Steven Weinberg, co-creators of the Standard Model, have explored alternatives to quantum mechanics. Weinberg eventually gave up when the alternatives he explored didn't work. 't Hooft, however, remains convinced that he can find an alternative, and he has a good track record.

So, the final known unknown of the physical world, and perhaps the deepest one of all, is the question: is the world, at its most fundamental level, governed by quantum mechanics? The answer will ultimately impact the outstanding mysteries I have described earlier in this book. Is the effort to find a quantum theory of gravity misplaced? Perhaps the theory that should fall by the wayside isn't general relativity, but rather quantum mechanics.

If quantum mechanics is subsumed by some more fundamental dynamics at the smallest scales, then the final stages of black hole collapse, black hole evaporation, the apparent singularity of the Big Bang, and the quantum creation of universes, are all subjects that will need dramatic alteration.

In my book, *A Universe from Nothing*, I argued that quantum creation of universes from nothing is to be expected, and that the properties of such a universe will, if it survives 13.8 billion years old, inevitably resemble the properties of the universe in which we live. But I acknowledged there (though few critics seem to have gotten that far in the book) that simply having no space, no time, no particles, and no radiation, may not be fully *nothing*. What about the laws of physics? Did they antedate our universe? (I emphasize once again that here language fails us, since if time didn't exist, then "before" and "antedate" have no meaning.)

The key point about inflation, or multiverses, is that most of the physical laws we measure today may be environmental. That is, they may vary from universe to universe and have no true fundamental significance. So, the Standard Model we measure, and the three families of elementary particles, and the whole shebang, may just be a fortunate accident.

But all of the pictures only make sense, at least mathematically, in the context of a quantum reality, a nature in which quantum mechanics is fundamental. I speculated in *A Universe from Nothing* about whether quantum mechanics too could have come into existence when our universe did, but I frankly really had and still have no idea what that means, or at least no way to mathematically define such a possibility.

Fortunately, nature doesn't care about what we can understand or what we can currently define. What is, is, and as marvelous and crazy as our quantum reality may be, and even if almost all physicists now would probably bet that that is the way our universe truly works— and perhaps the way all universes must work—we have to admit we really don't know. What we can say with confidence, however, as I hope the previous chapters have illustrated, is that the imagination of nature far exceeds that of humans, so unless we keep probing it, letting experiments drive our understanding, the known unknowns will never change.

4

LIFE

What is life?
How did life originate?
Is DNA life unique?
Are we alone?
What is the future of life?

Everything has been figured out, except how to live.

—JEAN-PAUL SARTRE

Nothing in life is to be feared, it is only to be understood. Now is the time to understand more, so that we may fear less.

—MARIE CURIE

Life is a series of natural and spontaneous changes. Don't resist them—that only creates sorrow. Let reality be reality. Let things flow naturally forward in whatever way they like.

—LAO TZU

The truth is you don't know what is going to happen tomorrow. Life is a crazy ride, and nothing is guaranteed.

—EMINEM

A man who dares to waste one hour of time has not discovered the value of life.

—CHARLES DARWIN

How did an inanimate universe become animate? That question has been the cause of wars and the inspiration for artists, painters, writers, and of course, scientists.

More than any other physical observable, life appears akin to a miracle. For many people, perhaps most people alive in the world today, it still is. Yet, at the heart of science is the presumption that natural effects have natural causes. If we accept that life is subject to physical laws, we are obliged to then move from the sacred to the profane. Or at least to the natural.

This conversation, when attempting to address the known unknowns of the universe, would be colossally incomplete if it did not raise what—for most people—are the two most awe-inspiring mysteries of nature: life and consciousness. Hence, the final two sections of this book.

While both of these domains are typically in the purview of the biological sciences, nature doesn't divide itself along the lines of nineteenth century academic disciplines. The laws of biology are determined by the laws of chemistry, which in turn are determined by the laws of physics. Any fundamental understanding of life will ultimately reflect the workings of these laws as well.

This fact has inspired some outstanding physics minds in the twentieth century to ponder the problem of the nature and origin of life, as well as its possible robustness in the universe. Erwin Schrödinger, one of the fathers of quantum mechanics, wrote an influential book in 1946 entitled *What is Life?*, which inspired a young student, planning to be an ornithologist, to turn instead to genetics. That student, James Watson, later discovered the double helix nature of DNA as the basis for the genetic code of life. (His scientific collaborator, Francis Crick, was trained as a physicist.)

As it turns out, Schrödinger was influenced by Max Delbrück, a physicist who had done seminal work on fundamental physics interactions but later turned his attention to genetics. His work in 1935 on molecular genetics strongly impacted Schrödinger. That work on bacteria and viruses, which won Delbrück the Nobel Prize in Physiology or Medicine, was done while he still had a teaching position in physics. He didn't become a professor of biology until 1947.

In the twenty-first century, the tools necessary to address the fundamental mysteries associated with the nature and origin of life are likely once again to be found in physics laboratories, and perhaps even astrophysics departments. The questions are too important and too fundamental to be relegated to merely one area of science.

Let's first consider Schrödinger's question:

What is life?

While it seems obvious and easy to tell if something is alive, upon more thought, the definition of life becomes particularly slippery. Ultimately, as Justice Steward said when referring to pornography, one seems driven to simply say, "I know it when I see it."

For example, one might posit the following definition: living systems reproduce faithfully and have an internal metabolism that draws energy from the environment and stores and expends that energy to grow and reproduce.

OK then—is fire alive?

It ticks all these boxes. A forest fire draws energy from the environment. It reproduces, and even somewhat faithfully, depending on the nature of the fire and the availability of combustibles. It certainly has a metabolism, expending energy to grow and reproduce.

But I don't think anyone would argue a fire is alive, so we need to do better. Here is a definition from Wikipedia: *Living organisms are generally thought to be open systems that maintain homeostasis, are composed of cells, have a life cycle, undergo metabolism, can grow, adapt to their environment, respond to stimuli, reproduce and evolve.*

This definition is definitely more complete and encompasses living things more closely. Homeostasis refers to the need to maintain biological equilibrium and was first described by the French physiologist

Claude Bernard in 1849. In 1920, Walter Bradford Cannon coined the term to describe this biological necessity.

The homeostasis requirement probably rules out fire since fire really doesn't have a moderating feedback loop to maintain a kind of static equilibrium. Nevertheless, it is quite reasonable to think of life, at least life that respires, as controlled burning. One of the surprising aspects of earth is that there was no free oxygen early on in the history of the planet. This was fortunate, because the process of oxidation releases energy, just as oxygen is needed to fuel most fires. Had oxygen existed early on, many of life's raw materials would have oxidized early on, releasing valuable stored energy necessary for the life that now covers our planet to begin, evolve, and grow.

Life exists somewhere between oxidation and fire. Unlike either, it moderates the intake of energy to maintain homeostasis, which is so crucial for survival, and ultimately reproduction.

But, speaking of reproduction, is this essential? What about viruses, like SARS-CoV-2, which has controlled all of our lives over the past few years? Viruses like SARS cannot reproduce on their own. They need to hijack the genetic machinery of other living cells to do so.

While they do not fulfill all the requirements to be alive according to the earlier definition, viruses definitely seem alive to me. They have a strategy that requires them to piggyback on other organisms, but they have hardwired the complex biochemical machinery necessary to reproduce in the environs of other living organisms. Moreover, I don't like the idea that I have to wear masks and get vaccine injections to protect myself against an inanimate object. So, I say viruses may as well be alive. But I also say that Pluto is a planet...

Alive or not, it is quite possible that viruses helped life evolve into its present form. Some viruses do not have a deleterious effect on their hosts, in which case, we can consider them as symbionts. Ultimately their biochemical features may be incorporated in more complex cells to allow living systems to expand their capabilities.

Perhaps the most famous form of such merging is mitochondria. These are the parts of modern living cells that control the intake and processing of oxygen in respiration. As Lynn Margulis and others first postulated, it is quite likely that mitochondria were autonomous organisms that were assimilated into other cells, enhancing their ability to process energy (respiration allows more than thirty-five times more energy to be released from the processing of electrons than photosynthesis does).

This is not unlike the more sophisticated assimilation of the Borg collective, for Star Trek fans. The Borg are an advanced civilization that conquer other civilizations, adapting and utilizing the best features of those civilizations into their own biology and technology. The first Eukaryotic cell that engulfed a mitochondrion didn't have the capacity to say, "Resistance is futile," but it probably was.

Definitions are useful, but they are not the heart of science (although unfortunately that is the impression that is given all too often in elementary science classes). Science is about processes, about understanding dynamics—and that is what I want to focus on. And while the study of the evolution of diverse life on earth is a rich and exciting field, involving its own puzzles, this is not where the dominant outstanding questions about life really lie. The big unanswered questions remain: How did life first began? Is life on earth unique? Is all life like life on earth? These are the questions I want to discuss here.

* * *

The first thing that jumps out at anyone who seriously considers the question of the origin of life is that even the simplest forms of life we see today are incredibly complex biochemical machines. It is clear that one cannot jump from a non-living world to a living world in a single step.

Some individuals who refuse to accept the likelihood of the natural formation of life and demand a supernatural explanation point to this complexity as proof of the inadequacy of natural mechanisms to

explain life's appearance. They use more modern examples of arguments presented by William Paley and others, saying that the claim
that RNA or DNA evolved naturally is like expecting a tornado to
blow through a junkyard and leave a complete Boeing 747 in its wake.

The problem with this viewpoint is that, just like evolution, the
origin of life was likely not a completely random process. The laws of
chemistry depend on the subtle interplay between the physical concepts of entropy—roughly, the tendency for closed systems to become
more disordered—and enthalpy—roughly, the stored energy allowing systems to do work at some temperature. Chemical reactions are
driven in certain directions depending on external conditions.

A good example of this, discussed by Schrödinger in his *What
is Life?* book, is the simple process of diffusion. Locally, seemingly
random processes nevertheless produce a certain directionality. An
ink drop in water will diffuse outward and inevitably uniformly color
the liquid. The opposite never happens.

It has been shown that under certain external conditions, for
example, it becomes thermodynamically and chemically favored for
elementary organic non-living systems to build up into more complex
molecules, the opposite of what we expect under the conditions we
normally experience.

This takes us back to an objection to the natural evolution of life
on earth that has been raised so often—especially by biblical apolo-
gists—that even Schrödinger felt the need to address it directly in his
book. The second law of thermodynamics appears to argue that disorder will inevitably follow from order, yet life is an example of order
following out of disorder. I get regular emails from people who raise
this point, as a kind of "gotcha" moment, expecting to demolish those
who of us see no need for supernatural intrusion into the origin of life.

The second law actually says nothing of the sort. It argues that in a
closed system—a system that does not exchange heat and energy with
its surroundings—the entropy (or disorder) of a large enough system
will not decrease over time. It can remain constant through processes

that are called adiabatic, or it can increase. However, all bets are off for *open* systems, namely those that exchange heat and energy, and sometime particles, with their surroundings. For such systems, local order can increase *at the expense* of the environment. This is generally manifested by heating up one's surroundings.

Of course, this is exactly what occurs with life, whose existence depends on extracting energy from its surroundings. In return, life returns heat, as well as other waste products. Each of us, at rest, is an 80-watt heater. If you want to see the bulk effect of this, go into a crowded movie theatre. When you arrive, it may seem chilly. By the time you and the crowd depart, it won't be.

This kind of local ordering is not unique to life, and in fact is so ubiquitous I am often surprised that people seem to ignore it when making their arguments about life. I am writing this in Canada in the winter, and one of the clearest examples is a snowflake. Viewed under a magnifying glass or microscope, snowflakes are beautiful renderings of the hidden order of nature. Just the electromagnetic interactions of dipole water molecules, acting under rapid cooling, form symmetric objects of wonder. They look like Christmas ornaments, and if we didn't know better, we'd assume they were carefully designed by a master artist. But, snowflakes form due to natural processes as energy is released when the ice crystals relax to the minimum energy configurations while temperatures quickly fall, releasing heat to their surroundings in the process.

You don't need to wait for it to snow, however, to see my point. Just look up in the sky on a sunny day. The sun is a beautiful ball of shining light. It has been able to persist against ultimate gravitational collapse for five billion years because of the immense amount of heat released into space, allowing it to continue in its ordered spherical form in spite of the chaotic and intense processes associated with nuclear burning at its core. In doing so, it provides the constant source of energy that allows life to locally fend off the global tendency toward disorder.

While the evolution of life from non-life doesn't violate any laws of physics, there still remain significant challenges. Returning to my earlier statement, even the simplest forms of life that have been discovered are incredibly complex. It is not feasible to imagine that these complex systems themselves arose spontaneously in their current form. There had to be precursors.

How we meet the challenges to the origin of life depends on what fundamental precursors you think are necessary for life to evolve.

The Greeks, as you may recall, imagined four basic substances: Air, Earth, Fire, and Water. For centuries, a debate ensued about which of these might be truly primordial—the basis of the others. The back-and-forth resembled the game Rock, Paper, Scissors. Each time one substance was proposed, good arguments could be presented for why it was trumped by one of the others.

A similar debate ensued among origins of life theorists. There are four key components associated with observed life on earth: informational molecules/genomes to enable faithful reproduction; the building blocks of metabolism, typically molecules called ATP that allow life forms to store and manipulate energy; protein catalysts to allow biological reactions to proceed; and compartments/membranes that separate the workings of living things from their environment. Each of these could reasonably be considered fundamental, and different research groups have explored different choices. Yet, at the same time they are all interconnected, so it is difficult to imagine the utility of one without the other.

The interconnections have pointed at another approach, which has been spearheaded by John Sutherland and his colleagues at Cambridge. One of the many misconceptions promulgated by Intelligent Design advocates, who argue against Darwinian evolution, is that the function of various biological components has remained constant over time. An object called the bacterial flagellum, basically a tiny bacterial motor, appears so complicated that advocates of teaching of Intelligent Design in high school science classes (including those who supported the defendants in the *Kitzmiller v Dover School District* trial) argued that it had to have been designed, namely that it couldn't have evolved by natural selection. But biologists showed that key components of the flagellum appeared in earlier lifeforms with different functionalities, so that the flagellum didn't evolve at once with its final purpose, but rather evolved over a series of intermediate steps, with various traits being selected for different functionalities. Similar arguments apply for the eye and visual system.

Analogously, when considering prebiotic chemistry, it may be a mistake to consider these four components of living things as arising separately and independently, and that the precursors for each of them had to arise independently of the others. It could be that intermediates exist with discarded traits that were essential to the development of these different elements simultaneously, perhaps with different functionalities at the time.

The interconnections between the different components, which initially appeared to complicate the effort to look for one precursor, actually suggest that these different components had common intermediates. For example, the biological machine that makes RNA is made from proteins, but the biological machine that makes proteins is made from RNA. Similarly, membranes, composed primarily of substances called lipids, are also composed of proteins that are crucial to metabolism. And metabolism is controlled by RNA and proteins.

Indeed, one of the differences between living systems and fire is that life is sustainable because it can turn off reactions. It is "slow-burning" rather than just burning. And if the reactions that govern living systems occurred naturally and easily and didn't require protein catalysts to assist them, you couldn't turn them off. They would be uncontrolled, and, like fire, would eventually burn through all the available fuel.

This fact means that it is not easy to see how prebiotic synthesis of the ingredients of life could easily work on the early earth, given the chicken-or-egg dilemma in that most biological processes require protein catalysis but the proteins necessary for the catalysis are synthesized by the biological processes.

It is here that significant progress has been made, and it is here that non-intuitive surprises have pushed the field forward over the past decade or so. Under extreme or unusual conditions, what seem like disfavored or unlikely reactions can become favored and can naturally occur. Sutherland's group, for example, has found that in extreme conditions where meteorite impacts might have occurred, and where there is ample UV radiation as well, exotic chemistry can utilize identical intermediaries to create vastly different biological molecule end products. They have also found how rather toxic solutions in primordial rivers and lakes that flowed over rocks could combine to produce a fertile set of organic building blocks. They have explored how exposure to UV light can assist to create many of the components of one of the most important biological metabolism cycles, the Krebs cycle, which produces energy-rich ATP molecules. Finally, they have found a suggestive path by which prebiotic synthetic chemistry might be then coaxed into producing the same end products via the utilization of enzymes, which would represent a nascent stage of biology.

This work demonstrates how extreme environments—utilizing organic chemistry that is sophisticated, complex, *and* non-intuitive— may have played a key role in the origin of life. Also, it shows how focusing on a broad set of initial materials and initial conditions, whose presence is suggested by geological and astronomical observations, at least plausibly allows a possible road map toward the emergence of the multiple components of life from non-life.

In some sense, this modern work hearkens back to more sophisticated and thoughtful versions of the seminal 1952 Miller-Urey experiment. There, when a mixture of gases that were then thought to be a part of the atmosphere of early earth—water, ammonia, methane, and hydrogen—was simply subjected to sparks mimicking lightning, it resulted in a rich organic mixture, including many amino acids, the key building blocks of proteins. While it is now thought that the early atmosphere didn't have this composition, it sparked the realization that complex precursor materials that could plausibly exist in the right environments could result in a surprising array of chemical reactions with extremely interesting end products.

As Jack Szostak, a Nobel Prize–winning geneticist who over the past decade or so has turned his research focus to origins of life questions, has emphasized, there are many startling chemical surprises that overturn conventional wisdom, if one is willing to let nature guide one's investigations.

For example, one of the big puzzles in biology (and again, one that has motivated Intelligent Design enthusiasts to propose life is impossible without design) is the fact that many biological molecules like amino acids have a handedness, and while they can come in right- or left-handed forms in general, life only exploits one-handedness. Biological molecules are generally left-handed. How could this arise in a background environment that doesn't distinguish between left- and right-handed molecules?

This has some vague qualitative resemblance to the puzzle of the matter-antimatter asymmetry of the universe, except that it is experimentally much more accessible, and also that the likely resolution is much less exotic.

Non-intuitively, starting with crystals of different chiralities (handedness) but with a distribution of sizes, one finds that smaller crystals, which are of one chirality, can preferentially dissolve in a solution that has equal left- and right-handed molecules. As a result, the other chirality can preferentially be accreted onto the larger crystals that haven't dissolved. Then over time, if this process repeats and crystals are ground down, an initially racemic mixture (a mixture with even left- and right-handed components) can evolve naturally into a solid state dominated by one chirality.

The process was so unexpected that people didn't believe it until the results were reproduced numerous times. Empirically, cosmic processes like these must have occurred at some point because some meteorites have been found with great anisotropies in the abundance of left- and right-handed amino acids. (Perhaps I should have added some time ago that amino acids have been found to be abundant in meteorites, so non-biological processes can clearly produce them without the need for the Miller-Urey "atmosphere.") This doesn't answer the question of exactly why life on earth exploits only left-handed amino acids, but it demonstrates how the early precursors of life could have arisen in an environment that was not racemic.

Another surprising example occurs in cold environments. Traditionally, to encourage chemical reactions to proceed more quickly, one heats up a sample. However, for nucleotides, the basic building blocks of nucleic acid-based polymers like DNA that can react to form long polymers (like DNA), sometimes reactions do not occur in solution. But if a sample is frozen, then after some time, polymers will have formed. This is completely non-intuitive. But it turns out that when the solution freezes, zones of pure water form, and there are regions with high concentrations of the nucleotide impurities at the boundaries of these regions. When these objects are highly concentrated, they are pushed so close together that reactions that would otherwise not take place are now able to occur. What makes this surprising result even more interesting is that one could imagine that this would be relevant to environments on the early earth or in space.

Szostak and his colleagues have focused on similarly non-intuitive environments in which chemical dynamics can thermodynamically cause individual ribonucleotides to bind together to build up long molecules like RNA, with single nucleotides being added one at a time when coupled to some initial templates. (Sutherland in his lab has focused on how one might synthesize the ribonucleotides in the first place).

The details are complicated and probably not illuminating to repeat here. The key point is that all of these unexpected chemical processes now make pathways where, in exotic environments, the natural synthesis of materials at the cusp of what we would call life might plausibly occur. Forty years ago, no such plausibility realistically existed. While the details of the origin of life remain among the great unknowns of science, through a series of baby steps—interspersed with periodic giant leaps over the past fifty years—it is now not unreasonable to think that this mystery may be resolved in the coming decades.

There is a lot of work to be done, of course, and I am reminded of the Sidney Harris cartoon where two scientists are looking at a blackboard with a long equation, at the center of which is written "then a miracle occurs," and one scientist says to the other, "I think you should be more explicit here." It is true that we don't have the specifics yet, but neither does it seem that a miracle is required.

You might wonder why Sutherland and Szostak have been focusing on RNA. That is because of one of the most surprising discoveries that initially suggested the question of the origin of life might actually be chemically solvable. It was so astounding that it was awarded the Nobel Prize.

This seminal development in thinking about the origin of life occurred when Sidney Altman and Thomas Cech, working independently in the 1970s, discovered that RNA (the molecule which transcribes the genetic code of DNA into a recipe for the production of proteins from amino acids, with the proteins then providing the catalysts that power and direct the chemistry of living systems) was also an enzyme that could catalyze reactions.

As I have described, most of the chemical reactions that power life require such enzyme catalysts in order for biological reactions to proceed as well as to be regulated and stopped. One of the chicken-or-egg problems that faced origins of life researchers early on was how to ultimately synthesize large molecules like RNA and eventually DNA without the protein enzymes that DNA instructs RNA to build in living systems. With the recognition that RNA could not only encode genetic information, but also catalyze reactions, the possibility arose for an "RNA world," a precursor to the world now dominated by

DNA and biologically synthesized proteins. If RNA could result from natural prebiotic chemistry, then this might provide both the genetic basis for reproduction as well as catalytic mechanisms for metabolism necessary for early life to eventually evolve into DNA-based life.

Thus, one of the oldest and deepest human questions—the origin of life on earth—which for most people is the deepest scientific question addressing our own existence, still remains unanswered. Nevertheless, tremendous advances have been made that suggest a scientific answer *is* possible, and both experimentally and theoretically accessible. The executive summary is as follows: (1) Exotic environments have been shown to exist where interesting new chemistry that can create complex biomolecules can be thermodynamically favored even without the preexistence of biological metabolism. (2) Many basic organic building blocks of life, including amino acids, cyanides, and other key molecules have been discovered in prebiotic systems, including meteorites and comets, and thus would have existed on a prebiotic earth. (3) The fact that RNA has both genomic and catalytic functions strongly points to a possible pathway to create replicating structures that also can catalyze key chemical reactions necessary to sustain these systems, in advance of biology, and can also guide the ultimate evolution of living systems with their own metabolism.

Several key associated questions remain. In order to answer how life originated on earth, we will need to understand where life originated. For a long time, the best bet was felt to be deep-sea underwater vents, with lots of available energy and highly reducing chemical environments. The recent work I have described, however, demonstrates the utility of ultraviolet light, and beyond that, possible extreme events like meteor impacts as possible sources of key biological building blocks. These suggest life may have begun on or near the surface of earth, as nascent continents rose up, and streams and lakes created possible breeding grounds for complex biomolecules.

Like much of science, the answer to these questions is most likely going to come from new experiments and observations that help guide researchers, often confronting what may now be conventional wisdom.

* * *

To paraphrase a line from Schrödinger, which, as we shall later see, he used in another context: life is a singular of which the plural is unknown. One of the biggest obstacles to understanding the origin of life on earth may be the fact that we currently have only the one example of life: DNA-based replication powered by ATP energy storage and generation. Without knowing the locus of possibilities, it is hard to distinguish specific pathways that may have been essential here on earth.

To move beyond this limitation, we can turn our eyes upward, in turn addressing that other great unknown: are we alone in the universe?

This question can be divided into two separate parts: is there life elsewhere in the universe, and is there intelligent life elsewhere in the universe? The two questions are profoundly different, and for many among the public, the second is the one of primary interest. For scientists, however, the first question is the one we really want to answer (not least because it is a precursor to addressing the second question), and moreover, is one we will likely be able to answer in this century.

For as long as humans have stared up at the red planet with telescopes, our planetary cousin, Mars, has beckoned us with hope. As early as 1719, the Italian astronomer Giacomo Filippo Maraldi noticed that the planet had seasons, which was confirmed sixty-three years later by William Herschel. Almost a century later, the French astronomer Emmanuel Liais postulated that the variations that were visible over the course of the Martian year were due to changing vegetation, thereby propagating the seemingly natural assumption that Mars also had life on it.

In 1877, the notion that there was not just life but *intelligent* life on Mars entered the scientific canon with full force. Astronomer

Giovanni Schiaparelli turned a new large refractor telescope in Milan toward Mars and perceived deep trenches crisscrossing the planet, which he called *canali*; the map he drew bears an uncanny resemblance to nearby Venice.

The translation of canali (trench) into canal was promoted by the American astronomer Percival Lowell, whose 1906 book *Mars and Its Canals* promoted in popular consciousness the notion that Martian civilization was more advanced than earth's, with a canal system that transported water from the planet's poles to its drier central plains.

While this false impression of the Martian surface was quickly debunked, the image of Martians persisted in the public imagination, reaching a new fervor when Orson Welles produced his radio drama *War of the Worlds* (based on the book written by H.G. Wells, which was motivated by an 1895 book about Mars by Lowell) which caused many to believe that earth was under an invasion by hideous Martian beasts.

After NASA's Viking lander showed pictures of the Martian surface that have now become familiar—with modern rovers crisscrossing the planet with their more advanced cameras, showing a dry, cratered, and red desert—Martian fever abated.

Nevertheless, moving from the world of science fiction and mistaken science to the world of modern science, Mars still remains the most exciting (and most accessible) candidate on which to find evidence for extinct, or extant, signs of life. Evidence of liquid water having flowed on its surface in the past, along with substantial evidence of water on or beneath its surface in places, has raised hopes of finding evidence for past or present life. Water is, after all, the source of life on earth.

But Mars has revealed something else that has changed our thoughts about life elsewhere in the cosmos. Those who are old enough may remember a press conference on August 7, 1996, hosted by the President of the United States, Bill Clinton. He described a meteorite found in Antarctica in 1984 and later analyzed by a team of chemists:

More than four billion years ago this piece of rock was formed as a part of the original crust of Mars.... It arrived in a meteor shower thirteen thousand years ago. And in 1984, an American scientist on an annual U.S. government mission to search for meteors on Antarctica picked it up and took it to be studied. Appropriately, it was the first rock to be picked up that year— rock number 84001. Today, rock 84001 speaks to us across all those billions of years and millions of miles. It speaks of the possibility of life. If this discovery is confirmed, it will surely be one of the most stunning insights into our universe that science has ever uncovered. Its implications are as far-reaching and awe-inspiring as can be imagined. Even as it promises answers to some of our oldest questions, it poses still others even more fundamental.

What caused Clinton to wax poetic and hold a press conference about a meteorite was the claim that within the rock were structures that bore an uncanny resemblance to the oldest fossils of life on earth. If true, it meant that life existed on Mars about the same time as it began on this planet.

What were claimed to be fossils in the Martian meteorite we now understand as being most likely due to non-biological processes; similar structures, having nothing to do with biology, have been observed on earth as well.

But what the saga of Allan Hills 84001 did was focus our minds on an important fact. The supposed fossil was not discovered on Mars, but on a Martian meteorite that fell on Antarctica. That, combined with the discovery of extremophiles—that is, life forms on earth that can exist in environments that would otherwise be considered impossible, from boiling hot springs, to acidic pools, to rocks deep underground—and the suggestion that single-cell life forms embedded in rocks could survive the journey from Mars to earth, meant that it might not be appropriate to consider Mars and earth as isolated eco-systems. In other words, life that evolved on Mars could have seeded the earth, or, somewhat more unlikely but not impossible, vice versa. If you wonder what Martians may look like, staring in the mirror might be one way to find out.

This issue is serious. Indeed, Andrew Knoll, a colleague and a distinguished geologist who is on the science teams that are exploring the Martian surface for signs of life, once told me that if we discover evidence for life on Mars, extant or extinct, the greatest surprise for him would be if we were *not* cousins.

This has significant consequences. If evidence of life on Mars is discovered and we can determine that those life forms are DNA-based with a cellular structure similar to some living cells on earth, we will not be able to conclude that this is evidence of a second, independent genesis of life in our solar system.

The reason that this issue is important is that if we could conclusively show that two such independent geneses occurred in a single solar system, it would imply that life is ubiquitous in the cosmos. The galaxy would be rife with it, which would in turn completely alter the estimate of the probability of intelligent life existing elsewhere.

And that means unless a new, clearly non-terrestrial form of ancient life is found on Mars, we may need to look elsewhere to address this profound question. Fortunately, we have a lot of new places to look. Our spacecraft exploring the solar system have now conclusively demonstrated that the most exciting places to look for life elsewhere may not be on other planets, but on the moons of those planets.

Because of tidal friction caused by their proximity to the gas giants Jupiter and Saturn (and other nearby moons), icy moons like Europa and Enceladus are now the focus of great interest. Both have now been determined to have deep oceans underneath their ice coverings, and geysers on Enceladus have been analyzed by the Cassini probe and found to contain water, ammonia, salt, and organic compounds.

Planning is underway for a mission to Europa to try and probe below the thick ice covering. Because of this covering on both these moons, the oceans below are truly isolated environments. Any evidence for life below their surfaces would give strong evidence of an independent genesis.

Other more exotic possibilities exist. Saturn's moon Titan, whose surface temperature is minus 180 degrees Celsius, has rivers of liquid ethane and methane. It is hard to imagine a less hospitable place, but the active chemistry on Titan is causing some to suggest it might harbor some exotic forms of life. NASA is planning a Titan probe as well, to follow up on the earlier Huygens probe which took images as it descended through its atmosphere.

Finally, our other close planetary neighbor, Venus, with hellish surface temperatures exceeding 450 degrees Celsius, was long considered a sterile planet. But there has been a resurgence of interest in possible life here too, in this case, in the clouds above Venus. These are so dense that there are regions in the clouds with earth-like pressures and temperatures. While recent claims of evidence for biomolecules in these clouds have been discredited, nevertheless, the question of whether there could be forever floating lifeforms on that planet remains open and interesting.

So much for our solar system. A vast new window on the rest of the galaxy was opened in 1995, initially by two groups, the team of Michel Mayor and Didier Queloz at the University of Geneva, and almost immediately thereafter by Geoff Marcy, then at San Francisco State University, and his colleague Paul Butler. As observers and experimentalists often do, they measured something that I would have argued was almost impossible.

When a planet like earth orbits around the sun, it tugs slightly on the sun, causing it to wobble back and forth. Because the sun is a three hundred thousand times the mass of the earth, the tug is small, causing the sun to move back and forth at a speed of about ten centimeters per second, about the speed that a baby crawls.

A larger planet, closer to the sun, would produce a bigger effect. But in our solar system the gas giants are all far from the sun. That is why the discovery of a planet orbiting the nearby star 51 Pegasi—first announced by Mayor and Queloz, confirmed shortly thereafter by Marcy and Butler, and followed within two months by Marcy and Butler's discovery of planets around both 47 Ursa Majoris and 70 Virginis—was so surprising. These stars had giant planets, as big as or bigger than Jupiter, orbiting at distances far closer than Mercury orbits the sun. For example, 51 Pegasi B, as the first planet is called, is about half of Jupiter's mass (although 50 percent larger in size), but it orbits its star in a period of only about four days, putting it ten times closer to its star than Mercury is from the sun.

Until these discoveries, this phenomenon was thought to be impossible based on observations in our solar system. Only rocky planets would be expected to form and survive so closely to their stars. Of course, once nature showed us otherwise, it didn't take long for astrophysicists to figure out that giant gas planets can form far from their stars, but due to gravitational perturbations can slowly migrate inwards. Since the discovery of 51 Pegasi B, almost fifteen hundred gas giant planets have been discovered orbiting their stars, with the closest having an orbital period of only eighteen hours! Our solar system, far from being typical, may rather be an anomaly.

We can't jump to this conclusion immediately, however, because of the selection effects governing these observations. The way that 51 Pegasi B and the early planets were discovered relied on measuring the tug from these planets on their stars, causing the stars to move at speeds of about fifty meters per second. While about five times faster than the fastest sprinter, this is nevertheless a remarkably slow speed when you consider its effect on the light emitted by stars. This motion causes the frequency of light emitted by the star to shift back and forth, due to the well-known Doppler effect. Compared to the speed of light, this motion is almost imperceptible, causing a periodic frequency shift of only about one part in ten million.

This is why I would have guessed that it would not be measurable. But as usual, I underestimated the skill and perseverance of observers. Nevertheless, the fact that spectrometers at the time could only measure periodic relative velocity shifts of tens of meters per second meant that planet searches were only sensitive to the effect of large planets on their host stars. Moreover, in order to disentangle signals from noise, the signal must be measured over many orbits. But if observing times are on the orders of years at most, then observing many orbits requires orbital periods of days or months at most. These two factors then imply that only large gas giants orbiting close to their stars would have been detected in these early efforts, which is precisely what was seen.

Gradually techniques improved, and Marcy and his team, who discovered seventy of the first hundred extrasolar planets, were able to discern the existence of multiple planets around stars, and the first gas giant at a distance from its star comparable to Jupiter's from our sun.

Marcy, with collaborators David Charbonneau and Timothy Brown, then pioneered another technique that has since blossomed to become the most promising method for planetary discovery used today. If one is lucky, the orbit of a planet around a distant star will bring it between the star and earth. During such a transit, it can block out a tiny fraction of the star's light, less than 1 percent on average. Once again, I would never have thought that this small variation in brightness would be observable, given all the other factors that can cause variability in the observed brightness of stars. But with enough careful data, and the periodicity of the effect (which includes both a diminution in the star's apparent brightness when the planet is in front of the star and a brief enhancement before the planet passes behind it, when it reflects starlight toward the earth), it can be observable.

Using these early results, NASA launched the Kepler mission, with a telescope in space that focused on a small region of the galaxy, continuously monitoring about one hundred fifty thousand stars looking for brightness variations. The result was spectacular. To date, over five thousand extrasolar planets have been observed around a host of different types of stars. Due in part once again to selection effects, the observations favor smaller stars and larger planets, but at this point the distribution of observed planets around stars spans sizes down to earth-like planets and orbital periods as long as an earth year.

These results confirm one a priori expectation among astrophysicists: almost every star in the galaxy is likely to have a planetary system around it, as the dust-like accretion disk around stars that accompanies the early phase of collapse into a star builds up by collisions to form planetary systems. However, almost every other bit of conventional wisdom has been dashed. Basically, anything not ruled out by the laws of physics can and does occur, with planets around small stars, large stars, and even collapsed stars that have undergone supernova explosions.

What is most exciting about the rash of planetary observations is the possibility of earth-like planets residing in what has become known as the "habitable zone" around their stars. This refers to that region where enough starlight is impingent upon the planet to allow liquid water on its surface, but not too much to evaporate it. A host of possible habitable planets has now been catalogued, with most of the planets—again due to selection effects—surrounding stars much smaller than our sun, so that they can exist closer to their stars but still not be inundated with stellar energy that would evaporate water on their surfaces.

Even the star closest to us beyond our sun, Proxima Centauri, a red dwarf star about four light years from earth and only about 12 percent as massive as our sun, appears to have a solar system of perhaps three or four planets, with one planet, Proxima Centauri b, lying in what is estimated to be the habitable zone of that star.

As excited as astrobiologists—a growing field of scientists who are attempting to explore the possibilities of life in and outside of our solar system, with, one must say, a highly variable set of scientific standards—may be about habitable zones, it is important to stress that the fact that a planet may exist in a star's habitable zone neither means it is habitable nor that it even has liquid water on its surface.

In the first place, we know from our own understanding of earth, which is clearly within the habitable zone of the sun, that the distribution of continents on earth has played a vital role in determining whether liquid oceans exist on its surface. For at least one period, and perhaps more, about 650–750 million years ago, as continents drifted across the earth's surface, the entire planet froze over in what has become known as "Snowball Earth." So even earth did not have liquid water on its surface at all times.

Beyond this, however, many of the planets within their star's habitable zones exist much closer to their host stars than the earth does to the sun, because the stars, like Proxima Centauri, are much smaller and cooler than the sun. These stars, however, are known to produce regular dramatic solar flares. Such flares could easily sterilize the surface of nearby planets. They could also be tidally locked to their stars, with the same side of the planet always facing the star, making for harsh conditions on both sides of the planets.

Thus, every time another earth-like planet is reported in the press, it is best to remain skeptical about the possibilities of finding life on it.

Nevertheless, given the proliferation of exoplanets, the odds of finding evidence for life outside of our solar system is growing every year. Moreover, we are quickly developing tools that might allow us to discover such evidence. With the launch on Christmas day in 2021 of the James Webb Space Telescope (JWST), the prospects have improved dramatically.

When a planet transits in front of its host star, telescopes on earth (or in the case of JWST, in space) can observe light that might pass through the atmosphere of the planet (if it has an atmosphere) on its way to earth. Observers can then try and observe how light from the star is absorbed by the atmosphere and also search for signals of radiation emanating from gases in the atmosphere.

The former effect, which allows spectral absorption lines associated with various compounds to be detected, has already been observed in this way for several extrasolar planets. The first such planet, HD 209458 b, has been observed to contain sodium, hydrogen, carbon, and oxygen, as well as water vapor. The atmospheres of other planets have included carbon monoxide as well as carbon dioxide, methane, and even exotic molecules like titanium oxide. As of this writing, JWST had already detected water and carbon dioxide on one giant planet, with the promise of much more.

Observing the different compounds making up the atmosphere of exoplanets will help determine if lifeforms like those on earth exist on these planets. For example, as I have described, there was no free oxygen in the atmosphere of early earth, and the oxygen content of our present atmosphere was produced by living systems over the past four billion years. Thus, detecting O_2 in the atmosphere of distant planets would be strongly suggestive, though not unambiguous evidence of life (as other non-biological mechanisms are possible). One can do better, of course, by examining a much wider range of potential biomarkers from methane to oxygen and beyond, and doing so might finally lead to a more definitive conclusion that life on earth is not unique.

The whole field heated up, literally, when JWST achieved full functionality in 2022. With its large size and infrared camera, the telescope can directly resolve the light emitted by hot planetary atmospheres, something not possible with terrestrial telescopes because the earth's atmosphere absorbs the infrared radiation. In addition, its resolving power is such that, with a coronagraph onboard to block out light from the host star, JWST can directly image these distant planets. It won't be able to resolve details, but it can provide images of the planets as points of light reflecting light from their host stars.

JWST has already reported its first results. These do not yet point to life outside of our solar system, but the potential exists for such a seminal discovery within the next decade. It is an exciting time.

I would be remiss if at this point if I did not at least mention an exciting, if wildly remote, possibility that involves a project I was involved in, funded by the Russian billionaire Yuri Milner. Milner has personally funded a number of cutting-edge science projects, at the level of about $100 million apiece, that involve new technologies that are speculative or which otherwise do not currently make the cut for significant government funding. These "breakthrough" projects include Breakthrough Listen and Breakthrough Starshot. The former extends the listening capacity of former SETI programs to listen to and watch 1M stars for signs of artificial radio or laser signals signifying the presence of intelligent civilizations. It is a long shot, but if you have the money...

Breakthrough Starshot, which was the program that I (and Stephen Hawking, among others) was a part of, makes Breakthrough Listen seem like a walk in the park. The idea is to use a bank of incredibly powerful lasers to accelerate a one square meter light-sail attached to a 1-gram spacecraft to 20 percent of the speed of light by the time it gets to a distance comparable to the earth-moon gap, at which point it could travel toward Proxima Centauri and pass by Proxima Centauri b in about twenty years, beaming back a close-up snapshot of the planet, which would then take four more years to be received on earth, if it could indeed be received. I won't go into the technical requirements, which are daunting, each one pushing well beyond current technological capacities. I went into the program feeling that it was worth exploring, even if it was a long shot, given that it was not costing the public anything. By the time I left the program, I felt it was far more in the realm of science fiction than I had first estimated. Nevertheless, if I am proved wrong, we might have a close-up snapshot of a potentially living world outside our solar system during the lifetime of at least my (not yet born) grandchildren.

* * *

Whether or not we discover signs of life elsewhere in our solar system or beyond—and I am personally quite confident we will (As Carl Sagan's protagonist in *Contact* once said, if we are alone in the universe is, it is an awful waste of space....)—what will really be most fascinating will be what *kind* of life we discover.

At least three possibilities exist.

First, extraterrestrial life might be based on exactly the same chemistry that we see on earth. While this may seem far-fetched, I don't think it is. As I have tried to describe in this chapter, the road to life likely followed a specific, if rare, chemical pathway driven by thermodynamics and energy considerations, as well as the availability of raw materials. Some aspects of life, like the four different nucleotide base pairs of DNA, might be random, so that perhaps life elsewhere uses DNA molecules with some substitutions for G, A, T, or C. But it could be that there are enthalpic or entropic reasons why only these combinations work. Similarly, RNA-like molecules in other lifeforms might code for different sets of amino acids. But once again, it could be that other combinations don't produce an effective array of protein catalysts. And it could be that life elsewhere is powered by something other than ATP molecules. But once again, it could be that no other molecule can be as easily synthesized, as easily manipulated, and as effective at storing energy.

In short, life could be *driven* by physics and chemistry to exactly the combination of components that life on earth discovered was workable and sustainable. I think this possibility is likely enough that I have made several bets with colleagues that any newly discovered lifeforms will mimic those seen on earth. It is a no-lose proposition as far as I am concerned, because even if I lose the bet, I win, because it means that life can be far more interesting and varied in the cosmos than the life we see here on earth.

A variation on this theme would arise in a possibility considered by numerous scientists but promoted most noticeably by the one of the discoverers of DNA, the brilliant polymath and scientist Francis

Crick. He was particularly fond of the possibility that has become known as panspermia. It is noted that life on earth arose about as soon as the laws of physics and chemistry allowed it, within about five hundred million years after the earth first formed, and after the early bombardment of the planet by potentially life-killing ocean-evaporating asteroids and comets began to subside. Perhaps the speed with which life began to take over the planet was not due to the ease with which the basic components were synthesized on the planet. Perhaps instead the seeds of life began elsewhere.

Our solar system is, after all, relatively young in a cosmic sense. The earth and sun are about 4.5 billion years old, whereas our galaxy is at least about twelve billion years old. Even if the stellar synthesis of heavy elements necessary for life required a few generations of early stars, there was still ample time for life like ours to arise elsewhere before life arose in our solar system. Perhaps the biomolecules necessary to kick-start life hitched a ride on interstellar grains, following a supernova that blew up a planet around a distant star or resulting from some mammoth cosmic impact. Or, perhaps, as some science fiction writers (and a few religious zealots, and some scientists) like to imagine, life was *intentionally* seeded on this planet by an advanced civilization.

As romantic as this notion is, like many such proposals, it really begs the question, and merely puts off the origins of life issue. If life on earth evolved because it was jump-started by life elsewhere, what about the life elsewhere? Was it seeded too? Eventually the buck will have to stop, and the non-biological origin of life will have to be explained. For this reason, I don't really take panspermia seriously.

The next possibility, which follows if I lose the bet, is that we discover organic life that has different rules, different genetic backbones than RNA and DNA, different metabolic pathways, and different sources of energy. Any of these could be the case if life is sufficiently robust to allow such changes, and one might expect such possibilities on planets or moons in which different combinations of raw materials are present—maybe less carbon or oxygen, maybe more silicon or nitrogen. It is well known that different supernova explosions produce different ratios of heavy elements. Our solar system was sparked by a supernova some five billion years ago. Life on a distant star system may derive from a different cosmic pool of ingredients, and if life is indeed ubiquitous in the cosmos it probably means that many different options are possible to choose from.

Finally, for me at least, the most exciting possibility is life that bears no resemblance whatsoever to life on earth, like the rock-like Horta on Star Trek. It is true that as we search for signs of life elsewhere in the universe, we are searching for conditions that mimic the conditions that led to the evolution of life on our planet as closely as possible. But this is because, like the drunk who loses his wallet, we are looking under the lamp post—not because it is the most likely option, but because it is the easiest hope for finding it. We know about life on earth, and we don't know of any other possibilities. As a result, it makes sense to explore similar systems first. If we don't find evidence for life there, then we can move on to more exotic options. And to quote Carl Sagan again, absence of evidence isn't necessarily evidence of absence.

* * *

The fact that we have no idea at all about the possibilities for life is exciting, but it should also temper our confidence in making estimates about the probabilities for life in the universe, particularly intelligent life, and also about the future of life in the universe.

In 1961, the astronomer Frank Drake wrote down what is essentially a mnemonic to guide thinking in order to estimate the number of active, intelligent, and communicative civilizations in the Milky Way. It has since become known as the Drake Equation, although it is not really an equation, but rather a back-of-the-envelope expression of the known unknowns. An estimate for the number of communicative civilizations is given as the product of a series of probabilities, almost none of which we have any empirical handle on. One of those probabilities was the fraction of stars that have planets around them, which we now know is approximate unity. The rest—the fraction of planets that can potentially develop life, the fraction that then do develop life, the fraction that develop intelligent life, the fraction of intelligent civilizations that develop communicative technologies, and the fraction that then attempt to communicate—were essentially just wild guesses at the time, and to a large extent still are.

We may never have an empirical handle on all of these probabilities, but we can pin some down. JWST, for example, may reveal the signatures for living systems on some planets, which would, for the first time, tell us that one of the key probabilities in Drake's estimate isn't near zero. The rest will depend upon learning more about the origin of life than we currently know. It is possible that chemists will determine the likely prebiotic pathway to an RNA world, or something like it, here on earth. But, as I indicated earlier, I suspect that the discovery of other geneses of life elsewhere may be required in order to confirm such a pathway, or to get a handle on the full variety of possibilities. If there are many different routes to life, these will undoubtedly include paths we would not otherwise suspect, because of our limited experience here on earth.

If organic lifeforms with fundamentally different biologies are discovered in our solar system, that will tremendously increase the odds that life is flourishing throughout our galaxy. And if new extraterrestrial extremophiles are discovered, such as lifeforms on Titan, then the possible locales for life to exist in the galaxy will be further multiplied. Our emerging knowledge about life on earth suggests that living systems here have colonized essentially every possible niche on the planet. Perhaps the same is true writ large for the galaxy.

In any case, as I have stressed, the field of astrobiology is still in its infancy, so that most claims one reads in the press should be considered skeptically. There is more speculation than there is data—perhaps well-informed speculation, but speculation nevertheless. Successful science is empirically based, and when it comes to the possible varieties and frequency of life in the universe, that empirical guidance is currently largely absent.

Having said all this, there are reasons for optimism. There are one hundred billion stars in our galaxy, and perhaps one hundred billion associated star systems, and life on earth evolved about as early as the laws of physics and chemistry would allow. Furthermore, water, organic materials, and starlight—the necessary ingredients for life on earth—are also ubiquitous in interplanetary space. I thus find it highly implausible that earth is the only life-supporting planet in our galaxy, much less the one hundred billion galaxies in the observable universe.

Intelligence is another matter, however. Intelligent life at the level of modern humans took almost four billion years to evolve on earth. There is no evidence that intelligence is an evolutionary imperative. Our own evolution required a series of fortuitous circumstances. Not least is the fact that earth is located in a remote suburb of the galaxy where no catastrophic galactic events interfered with the process. Closer to home, Jupiter cleared out most of the potentially destructive asteroids and meteors that could have ended earthly evolution at any point. Other accidents of fate include the asteroid impact that appears to have ended the dinosaurs' domination of the earth, paving the way for mammals to evolve as they have. Was all of this necessary for human-level cognition to have evolved? What other currently poorly understood accidents may have played a key role in allowing humans to evolve and survive?

Our lack of understanding of the varieties of possible intelligence, a topic relevant to the next section of this book, clearly impinges on the degree to which we can address these questions.

Finally, it has long been recognized, based on the history of the last century and by recent events, that intelligent technological civilizations may have a lifetime that is rather short in cosmic terms. Are technological civilizations doomed to self-annihilate before they communicate, or even venture out across the chasms of interstellar space? It remains to be seen.

* * *

All of these considerations come to play when considering one last set of unknowns having to do with the future of life. More specifically: Can life on earth survive the dynamic evolution of the earth and our solar system? If so, in what form? Beyond that, is the future of life tied to the ultimate future of the universe? And can life survive forever in an eternally expanding universe?

Independent of what modern humans are doing to the global climate in this century, and what geopolitical impact that may have, over the long run, life on earth will be governed by the dynamics of our sun and galaxy. On the surface, it doesn't look good.

Over time, the sun is getting brighter. During the early evolution of life on earth the sun was about 15 percent less bright than it is now. Had it not been for the fact that the carbon dioxide abundance in the atmosphere at that time was orders of magnitude greater than it is now, the earth would have been frozen over.

In two billion years, the sun will be 15 percent brighter still. With that level of radiance, earth will exist in a zone comparable to that currently occupied by Venus. Without intervention, there will be a runaway greenhouse effect, and temperatures on the earth's surface could approach those of Venus today, about 450 degrees Celsius— enough to sterilize the surface of the planet.

About five billion years after that, two combined astrophysical challenges will come into play. The least dramatic will be the collision of the nearby Andromeda galaxy with the Milky Way. While that may sound apocryphal, it isn't. Much of the galaxy is empty space, and the collision of stars with other stars is likely to be minimal. But the global gravitational evolution of the galaxy will change dramatically. Our spiral galaxy will be deformed into a spherical or elliptical galaxy. That alone may have little impact, but the possible close approach of nearby star systems could gravitationally perturb our solar system, and because our solar system's dynamics are actually chaotic, such small perturbations could have large results, including the ejection of planets from the solar system.

More dramatic, however, is the fact that in about five billion years the sun will exhaust its hydrogen fuel and evolve into a red giant star. As the core of the star collapses inward, becoming sufficiently dense so that helium will begin to undergo nuclear reactions to form carbon, the outer envelope of the sun will expand, encompassing what is the present orbit of the earth.

On the surface, this appears to spell complete doom for life on earth, and it will be, if nothing else happens. But, intelligent beings, if they survive in one form or another that long, have the capability of intervening in various ways. For example, by redirecting comets and asteroids to have close encounters with earth, gravitational energy can be exchanged, slowly moving earth's orbit outward. Over the course of a billion years or so, earth could move to an orbit closer to that of Mars, which will be far more hospitable at that time.

What we do to Mars is of course another question. As some have proposed, perhaps it would be easier to have a wholesale exodus of life from earth to Mars—rats leaving the sinking ship, so to speak.

As we begin to envisage these science fiction-like speculations, we can consider the broader question of whether life will hedge its bets and leave the solar system before these emerging potential disasters need to be confronted. Humans have evolved to function particularly well on earth. We are not so well-suited for interplanetary, much less interstellar, travel, beyond the obvious technical challenges of doing so.

If I were going to bet, my money would be on sending not humans but the instructions to make humans out beyond our solar system. Mass is the enemy of space travel, and human biology puts tremendous constraints on the possible forms of space travel. Witness today the difference in cost of sending a rover to explore Mars—something less than a billion dollars—to the cost of a round-trip human mission, perhaps $100 billion. With further miniaturization of command-and-control components, small spacecraft could be cheaply sent to the outer reaches of our galaxy.

It may seem harsh to consider sending small robotic systems on one-way missions to preserve our culture, knowledge, and perhaps even our biology, rather than sending humans themselves, but the logistics strongly favor the former over the latter. And who is to say the dominant intelligence on earth may not become silicon based, rather than carbon based, in the far future in any case?

Moreover, as many people have noted, sending humans beyond our solar system may require large, self-sustaining ships that move slowly, taking perhaps many millennia to cross interstellar distances, implying many generations must be spawned on the way there. By the time these beings arrive, in many respects, they may not closely resemble modern day humans either culturally or physically.

These ramblings may seem to have become too far removed from the more down-to-earth issues with which this chapter began. But instead of returning from whence we came, I will take things a few steps further to address a subject that goes by the name of eschatology: the doctrine of last things. Originally the domain of theology, like all good ideas, it moved into science, and it is the area in which I first began to do research before I had even heard of the term.

I learned the term, as well as many other things, from the brilliant physicist/mathematician Freeman Dyson, who achieved renown in physics for his seminal role in developing the quantum theory of electromagnetism. Freeman worked at the Institute for Advanced Study for over 60 years prior to his death in 2020. During this time, his fertile mind roamed over the vast expanse of human intellectual activity.

In 1979, for example, Freeman turned his considerable intellect to the abstract question: can life be eternal? That is, not an individual life, but rather a civilization. He removed himself from practical considerations (not uncommon for him) and addressed the more fundamental general physical question: how do the laws of physics constrain the future of life?

Ever the optimist, Dyson considered what seems to be the most optimistic case, an eternally expanding universe. A universe that ends in a big crunch seems less inviting. At the time, Dyson considered the possibility of an open universe as representative of the best option for eternal expansion. In the first place, at the time, the evidence suggested that the universe was most likely open. In the second place, the possibility that the expansion might be dominated by an energy in empty space, which, as I have described earlier, changes all the rules, was not taken seriously.

In his typical way, Dyson removed biology from the problem and thought about a generic lifeform experiencing time, processing information, and expending energy in the form of heat to its surroundings. He also assumed that any civilization had access to only a finite amount of energy, even over an infinitely long time.

Since it takes energy to process information, one might immediately think that such a civilization could not survive indefinitely. Dyson defined eternal survival for a civilization to be equivalent to the civilization having an infinite number of thoughts or conscious moments. The latter he defined to be the subjective time experienced by that lifeform, which he argued should sensibly be proportionate to the metabolism of the creatures, and hence their "body" temperature.

Then, in a typical bit of legerdemain, Dyson argued that life should adopt the following strategy: as the universe cools, life hibernates, then wakes up and thinks, and then goes back to sleep. Dyson was able to show that if the lifeforms sleep for a longer and longer fraction of the age of the universe as the universe ages, punctuating these periods of hibernation with brief moments of wakeful insight, then the lifeforms could experience an infinite subjective time while only expending a finite energy. Ta da! Eternal civilization. Of course, it should be noted that none of the practical details about how, and even if, this could be achieved were spelled out.

In the late 1990s I had been thinking about the consequences of the existence of dark energy for the future of the universe along with my colleague Glenn Starkman, and motivated in part by Dyson's 1979 article, we began to think about how his arguments would change in a dark energy-dominated universe.

It turned out that the addition of dark energy, which causes the universe to expand, destroyed Dyson's argument for very late times (technically this is because eventually the temperature of the universe doesn't decrease anymore, but approaches the so-called Hawking Temperature associated with an exponentially expanding spacetime). But more than that, we felt we had arguments that would counter Dyson's conclusions for any eternally expanding universe.

I contacted Freeman about this, and thus began one of the most enjoyable sparring dialogues I have ever had in physics. It turned out that I was going to spend the next term as a member of the Institute for Advanced Study in Princeton, where Dyson worked. We ate lunch together most days, but about once a week I would go into his office and present what I felt was a foolproof argument showing he was wrong. The next day he would come up with a brilliant counterargument. Ultimately, he agreed that in a dark energy-dominated universe, the persistence of life was essentially doomed in the long-term future, but he never gave in for the other cases. We left that discussion agreeing to disagree, and while we met numerous times after that, we focused on other topics.

One of the reasons I brought up this episode is that it was during this exchange that I learned from Dyson about an example of hypothetical life that had first been proposed in a science fiction novel by Fred Hoyle called *The Black Cloud*.

In the story, a cloud of dust-like particles is observed in space, and eventually researchers realize the cloud is a living entity, with the particles somehow communicating and combining to form a single consciousness with intent.

The story itself is not as significant as what Dyson did with it. Once again, without worrying about details, he pointed out that such a cloud, if it is allowed to expand with the universe, might represent just the kind of lifeform needed to obviate our concerns and allow a civilization based on such forms, or perhaps a single lifeform, to "live" forever.

I confess I had never thought of such disembodied objects when thinking about life, and while Glenn and I came up with arguments that we think destroyed Dyson's claim about this lifeform, his example should serve to remind us that we need to keep an open mind about life and what life means.

Since we don't know the possibilities, we need to realize that the laws of physics ultimately govern what is possible and that biology conforms to those laws, not vice versa. Therefore, when imagining the spectrum of possibilities of existence, it makes sense to remove biology from the problem, because you can lose the forest for the trees. Dyson's thinking, which is typical of the way physicists like to approach problems, was reminiscent of the joke with which I began one of my earlier books, *Fear of Physics*, where I allude to a physicist who is asked to consult for a dairy farm and who walks up to the blackboard, draws a circle, and proudly announces: "Assume the cow is a sphere!" Surprisingly, as stupid as this statement sounds, there is a lot you can discern about cows without modeling them any more carefully than that.

When it comes to life in the universe, or even life in a multiverse, we need to realize that we may have just scratched the surface, and if we are trapped into imagining only life that plays by the rules we have learned, our thinking will be limited. And while even a "Black Cloud" may not live forever, it represents perhaps the most important new object lesson that thinking about life in the universe is likely to teach us: that there are more things in heaven and earth than are currently dreamt of in our imagination.

And finally, if thinking about the possible ultimate end of civilization in a cold dark eternally expanding universe is too depressing for you to contemplate, you may take solace in a key fact illustrated in that Woody Allen joke: eternity is a long time, especially near the end. While no civilization may be able to survive indefinitely, in a universe that expands *forever*, there may be local fluctuations that might, with unbelievably small *but non-zero* probabilities, cause momentary (in a cosmic sense) emergences of some new life forms. They too will die out, but the process may reoccur every now and then, forever. In that vague and improbable sense, life itself might never end in the universe. It just won't be the same life that survives....

* * *

The arguments I just presented are important in thinking about another perceived unknown regarding life: was it designed? I think it is important to at least bring up this topic because so many people seem to find it seductive.

I should state at the outset that there isn't a shred of evidence to suggest design in nature, and there is plenty of evidence against it. But, as I have previously stressed, absence of evidence isn't evidence of absence. Therefore, one cannot prove there is not a hidden grand designer. One can just argue that there appears to be no need for one. And one can debunk the many tired and false arguments that claim to show such a need.

The existence of the diversity of life on earth and its particularly spectacular fitness for its surroundings, was, for millennia, taken as proof of the need for a creator. All that changed with Charles Darwin and Alfred Russel Wallace, who demonstrated that natural selection, along with standard biology, could naturally result not only in diverse species, but species that appeared to be providentially designed for their surroundings.

In other words, they demonstrated that the earth wasn't fine-tuned to fit life, but that only life that was fine-tuned (by evolution) to fit the earth would survive.

I emphasize this point because for some reason the same debate has surfaced again, but this time in cosmology. The point is stressed over and over again by people who believe that there must have been an intelligent creator for our universe, that if any one of a number of fundamental constants took values that were even slightly different than their actual values, life as we know it could never have evolved.

This statement on its own is not incorrect. And it took on a new higher profile when a non-zero value of the energy of empty space (dark energy) was discovered, and its value was 120 orders of magnitude smaller than one might naively expect it to be on the basis of particle physics arguments. In other words, it seemed impossibly small. If it were even an order of magnitude or so larger, then galaxies would probably not have formed, and without galaxies there would be no stars, and without stars, no planets, and without planets, no people...

The conclusion drawn by some, on the basis of this fact, is that the fundamental constants of nature were, of necessity, pre-tuned so that we might be here today. What more evidence could there be for a divine planner?

There are many things wrong with this argument, however. Not least is the fact that if the energy of empty space was precisely zero, a value that most physicists had thought was a natural expectation, then the universe would be *more* fit for life over the long term than it is now.

More important however, are the considerations outlined in the previous discussion. It is true that if the parameters of the universe were different, we might not be here, but since we have no idea what the complete set of possibilities are for life, especially what the possibilities would be if the laws of physics were slightly different, then who are we to say that there wouldn't be some different kind of life that could arise in such a universe? A black cloud perhaps? And I expect in such a universe these lifeforms would be wondering why their universe was so fine-tuned for *their* existence!

The point is that the universe isn't fine-tuned for life. Rather, life on earth arose because it could. Just as in the case of biological evolution, life is fine-tuned for the universe, rather than the other way around.

The existence of life in our universe seems miraculous, but it need not be a miracle. The mysteries surrounding the origin of life, its variety, and its possible future are fascinating and provocative. The fact that we don't yet fully understand these things is not evidence for God or that we live in some vast video game created by some more advanced civilization (which of course begs the question of whether they live in a video game, and so on). Rather, it is simply evidence of not understanding, and that motivates trying to find out the answers. Hoyle's black cloud, and the considerations that stem from it, should give us some cosmic humility. Not only are we not likely to be special gifts of creation, but the possible existence of lifeforms with nothing in common with us lends further incredulity to the incredibly non-humble suggestion that the universe was made for us.

5

CONSCIOUSNESS

What is consciousness?

How does it arise?

Are humans the only conscious animals?

How will consciousness evolve?

Can we create it?

Should we create it?

*Consciousness is a singular of which
the plural is unknown.*

—ERWIN SCHRÖDINGER

*No problem can be solved from the same level
of consciousness that created it.*

—ALBERT EINSTEIN

*Consciousness is much more than the thorn,
it is the dagger in the flesh.*

—EMIL CIORAN

*What a piece of work is man, how noble in reason,
how infinite in faculties, in form and moving, how
express and admirable in action, how like an angel in
apprehension, how like a god!*

—WILLIAM SHAKESPEARE

*And my head I'd be a scratchin'
While my thoughts are busy hatchin'
If I only had a brain.*

—HAROLD ARLEN/YIP HARBURG

When I was young, I quickly became entranced by science, by the possibilities of the universe, and by the possibilities of being involved in making discoveries about the universe. Being the first person in the world to understand something fundamental felt, for me, like the greatest adventure anyone could ever hope to have.

The first field I focused on was what today would be called neuroscience. For me, nothing seemed like a greater challenge than understanding the brain. In many ways, it still does.

I didn't think about neuroscience per se, because I had never heard of that term or that field. My mother had big plans for me and my brother. He would be a lawyer, and I would be a doctor. He did become a lawyer. My plan was to become a brain surgeon, because I thought a doctor was as close to being a scientist as anyone could be.

Needless to say, I never became a doctor, and I also didn't become a neuroscientist (though I toyed with it from time to time when depressed in graduate school). What neuroscience filtered down to me seemed like taxonomy, discovering what the parts of the brain were, but not really understanding exactly how any of them worked and, more importantly, how they worked together to produce thought. It just seemed hopelessly complicated. And once again, in many ways, it still does.

Nevertheless, as with many fields of science that I chose not to study, the past half century has seen revolutionary changes in our understanding of the brain, based largely on the development of new tools to probe the brain. We can probe single neurons, observe a thinking brain in action with functional magnetic resonance imaging (MRIs), read out data that in certain cases allow us to predict what a person is seeing or even thinking about, explore the perceptions of individuals with known brain damage, and even interface human brains with mechanical devices to try and allow individuals to move limbs they previously couldn't move.

Yet the mysteries that captivated me, and I suspect captivate all of us, still remain largely unresolved. What makes us *us*? How do we come to model the world and ourselves within it, predict it with foresight, and reflect upon it and our own existence afterwards? What makes us self-aware? Are we the only animals who are? There is perhaps no deeper personal mystery in the universe, yet perhaps no subject presents more obstacles to human exploration.

I once heard that the amount that is known about a subject is inversely proportional to the number of books written about it. That probably explains why there are so many books about consciousness coming out at an astonishing rate. I have been fortunate to have had long conversations with a number of neuroscientist-authors and have explored numerous books and articles on the subject. I continue to be struck by the fact that different authors claim to have viable theories of consciousness, but each one differs in some way from others. Just like different religions, they can't be all right, and perhaps none of them are.

On the other hand, new tools have resulted in a great deal of progress in neuroscience, psychology, and medicine over the past half century. The different perspectives presented by neuroscientists are illuminating because they reflect how different selection strategies from amongst the growing wealth of data in this field point toward different possible solutions of what the Australian philosopher David Chalmers coined "the hard problem of consciousness"—how physical processes in the brain associated with storing and processing sensory data through biochemical reactions produce a mental image of the world and of ourselves that governs how each of us live our lives.

That Chalmers is a leading scholar in consciousness studies reflects a fact about the field that might give one pause. In stating this, I am virtually certain that it will be taken the wrong way, as I already have a reputation for dissing philosophy, in spite of the fact that I value it as an important human intellectual activity. That said, the phenomenon of consciousness is the one area I know of in science where the forefront discussions seem to be made by philosophers equally as

166

often as they are made by experimental cognitive scientists. I tend to view this as an indication of a science in its early stages. Philosophy is indispensable for developing the questions in fields of science where it is not clear what the important questions are, namely, the ones that scientists will use later to explore nature, which in turn lead to further questions, answers, and so on.

In a field like physics, part of what was originally called natural philosophy—the early philosophical questions about motion, preferred states of being, and so forth—gave way to specific mathematical questions that were once removed from the original philosophical issues, following the empirical efforts of Galileo and later the theoretical mathematical insights of Newton.

Over time, physics progressed greatly, and the questions that occupy almost all physicists today bear little or no direct relationship to the issues discussed by philosophers of science today. Without being pejorative, it is a simple fact that physicists on the whole don't read philosophical literature. The two areas have largely diverged, simply because physics has made so much progress as an empirically based science.

But with neuroscience and the issue of consciousness in particular, things are quite different. It remains a subject of ongoing debate to decide what are the fundamental questions or concepts that will most likely produce progress. And so, philosophers can usefully parse logical issues and propose concepts that can help direct future research. Philosophers like Patricia Churchland and Daniel Dennett, having steeped themselves in science issues, seem to play a significant role in pushing debate and research forward, as do neuroscientists such as Antonio Damasio, Joseph LeDoux, V. S. Ramachandran, and the late scientist Francis Crick, and cognitive scientists like Steven Pinker and Noam Chomsky, to name some scientists who have influenced my own understanding of these issues. (Some of my other influences include popular writers such as Susan Blackmore, Anil Ananthaswamy, and of course, Oliver Sacks.) This variety alone provides empirical evidence, as I perceive it, that the key questions of consciousness are far from being resolved.

Here's another problem. Trying to define life, as I described at the beginning of the last chapter, is challenging. But that is nothing compared to trying to define consciousness. The science of consciousness is so fluid that one can find almost as many different definitions as there are researchers. In fact, many scientists who focus on consciousness steer away from a strict definition, working to develop an understanding of what most importantly captures the nature of consciousness by inching toward human consciousness through an evolutionary consideration of neural processes more generally, or by exploring what might be wrong or missing from various conventional definitions.

Consciousness is a slippery quality because it exists on a spectrum in the evolutionary development of life that is very difficult to measure or quantify. Unlike legs, or fins, or an eye, one cannot objectively measure cognitive development in a long-extinct species of animal. And even in extant species, there are no direct probes of what one might call consciousness. This is because many key features of conscious self-awareness, determined by observing behavior, are possessed by lifeforms without brains, which are clearly not conscious.

As a result, one has to go beyond observations of behavior in order to probe the depths of consciousness, but only for humans do we even have the possibility of reporting on direct self-observation. For this reason, there remain significant debates among different neuroscientists on fundamental questions such as the level of consciousness of animals like dogs or cats, not to mention octopi, dolphins, and whales. To me, my dog appears to clearly have memory, emotions, and rudimentary reasoning skills. Is he conscious? It certainly feels like it when I look in his eyes. Am I anthropomorphizing here? Of course. But am I wrong?

Bertrand Russell once remarked, regarding the interpretation of animal behavior, that "all the animals that have been carefully observed have behaved so as to confirm the philosophy in which the observer believed before his observations began."

Indeed, one of reasons why I remain skeptical of my perception about my dog's feelings is that we are all conditioned to ascribe behavior to feelings. But not only is this not always the case, it may be almost never the case. We see a rat freeze when a cat approaches and ascribe that freezing to fear, but evolutionary arguments, as I shall describe, suggest that it is actually a survival response that has as much in common with the musings of Descartes as it does with the motion of a bacterium. If animals have feelings, it is just as likely that these feelings result from behavior as vice versa.

Even with the ability to communicate with conscious subjects, such as humans, about their conscious experience, one should remain skeptical about the results of interrogating subjects who relate the cognitive basis of their behavior. Most of us may imagine our behavior is based on rational decision-making, but in fact, we are often fooling ourselves. We are simply not conscious of all the factors that lead to the decisions to speak or act. We then invent plausible explanations for ourselves, which go more familiarly by the term "rationalization." If we don't fundamentally know why our conscious minds are causing us to act, we certainly cannot convey that information to experimentalists who may be studying us.

One of the most convincing demonstrations that we can fool ourselves arises from the so-called split-brain experiments devised by Roger Sperry and Michael Gazzaniga in the 1960s.

Our brains have two hemispheres. They are primarily connected by a region called the corpus callosum (cc). Men have smaller cc's than women, which is why my wife claims I am much worse at multitasking than she is. Be that as it may, it is well-known that some sensing information crosses from one hemisphere to the other. For example, visual information from the left side of your visual field is processed in your right hemisphere, and vice versa, whereas auditory information from the right ear is processed in the right hemisphere (and similarly, left to left). Motor control, like visual information, switches hemispheres, so that the left half of the body is controlled by the right hemisphere and vice versa.

As long as the connections between the two hemispheres are normal, this division of labor is not really noticeable since information is processed quickly by both halves. In the 1950s and '60s, however, severely epileptic patients were treated by severing the cc in order to prevent seizures from spreading from one side of the body to the other. The results were encouraging, with seizures abating and few other noticeable changes in behavior, personality, and so forth. It was on these patients that Sperry and Gazzaniga performed their observations.

They would have patients sit in front of a divided screen and stare at its center. Words or pictures were flashed to one side of the visual field or the other. Patients could respond verbally or by using their hands. Speech processing is primarily restricted to the left hemisphere, so that if a picture was flashed on the right visual field the patient could describe it normally. If it were flashed on the left visual field, however, the patient could not.

For objects on the left visual field, patents could not verbally describe what they saw, but they were asked to describe what they saw without words, using their hands. Since the right hemisphere controls the left hand, patients were given a bag of objects and were asked to select, with their left hand, which object corresponded to what they had seen.

In one experiment, a patient was shown a snow scene on the left visual field and a chicken claw on the right and asked to pick out matching pictures from an array in front of the patient. With their left hand they chose a shovel (corresponding to the visual image of snow their right hemisphere had processed), and with their right hand they chose a chicken (coming from the right visual field processed by the left hemisphere).

So far so good. But here is the kicker. When asked to verbally explain the reasons for the choice, with verbal ability being the province of the left hemisphere, the patient explained, "The chicken claw goes with the chicken, and you need a shovel to clean out the chicken shed."

In short, the verbal left brain created a coherent story of conscious awareness that had nothing to do with what the person actually saw. The brain rationalized the choices to explain behavior, communicating a false narrative because that was the only narrative that made sense for the left hemisphere.

Two different conclusions can be drawn from these observations, and the two experimenters differed in their choice. Sperry concluded there were two different consciousnesses experienced in the two hemispheres. Gazzaniga concluded that it is only the left hemisphere that really generates high level consciousness because that hemisphere controls language and beliefs and ascribes intentionality and actions to people.

While demonstrating the diversity of the viewpoints of researchers, even on the same data set related to consciousness, this also demonstrates the difficulty of trying to infer the nature of conscious reasoning in subjects by asking them questions or viewing their behavior, because they are likely to not relate what the actual source of that reasoning is, not out of intentional subterfuge, but simply because we cannot always access the true causes of our actions. As David Hume presciently observed in *A Treatise of Human Nature*, "Reason is, and ought only to be the slave of the passions."

This problem is reflected in Albert Einstein's quote at the beginning of this chapter. We are conscious beings trying to understand the very processes we use when trying to understand consciousness. While we can now probe neural processes in other individuals, we cannot really get inside their heads to know what they are thinking and how, without relying on their own reporting of such. Moreover, we cannot get outside our own heads when attempting to study ourselves. We can only experience personal perceptions of the end result of consciousness in ourselves and others, and not the process by which it comes about or how closely our perceptions match reality.

This fundamental obstacle was recognized explicitly by psychologists as early as the 1890s. In William James' classic book, *Principles of Psychology*, where he first coined the term "stream of consciousness," a concept we shall come back to, he spent significant time trying to look into his own mind for clear evidence of the origins of consciousness, and he argued such introspection was akin to "trying to turn up the gas to see how the darkness looks." Introspection at best takes us to the top of the iceberg of consciousness, but what is really going on is as hidden as the massive portion of an iceberg that looms below sea level. James even acknowledged that while perception itself might be simpler to understand than consciousness, "part of what we perceive comes through our senses from the object before us, another part (and it may be the larger part) always comes out of our own head."

George Mandler and William Kessen later echoed James' sentiments in their 1959 book, *The Language of Psychology*: "Atoms do not study atoms, stars do not investigate planets...The fact that man studies himself and that he has archaic notions which persist in the daily behavior of all men puts a major stumbling block in the path of scientific psychology."

The situation is reminiscent in some sense of our efforts to understand our universe on its largest scales and gravity on its smallest scales. On the one hand, we are stuck inside of our universe, and are thus limited in the kind of questions we can empirically explore; on the other, we do not yet have experimental probes that can differentiate between different theoretical predictions about the behavior of gravity on small scales.

In this regard, I was particularly stuck by a phrase by neuroscientist Antonio Damasio on the nature of consciousness: "...in discussing phenomena as complex as mental events are, we often have to settle for plausibility when verification is nowhere near." This is precisely how I have described events like the origin of our universe or the existence of a multiverse, where we can rejoice in arriving at plausible explanations, but hoping for verification or falsification at present is, once again, the stuff that only dreams are made of. Of course, in the case of consciousness, what dreams are made of is the very question of interest.

* * *

The eminent Ukrainian American evolutionary biologist Theodosius Dobzhansky wrote in 1973 that "nothing in biology makes sense except in the light of evolution." We are the product of every organism, extant or extinct, that has come before us. Whether or not consciousness first emerged in humans, whenever it did emerge, it did so through a long and winding evolutionary path that directly connects us to the first living organism on the planet.

Consider for a moment the minimal attributes one might think are necessary for consciousness to be manifest. A first guess might include behavior, learning, and memory. Here one immediately encounters a problem. While these are attributes are probably necessary conditions, they are far from sufficient.

To quote Russell again: "From the protozoa to man there is nowhere a very wide gap either in structure or in behavior." In order to pass on their genes, all organisms since the beginning of time have had two imperatives: to survive and to reproduce. To survive, even the most primitive lifeforms, bacteria and archaea, have needed to sense their surroundings, and move, if possible, to avoid harm.

Protozoa can swim away from harmful environments, including harsh chemicals and sunlight, and swim toward safer environments to balance their fluids or regulate their temperatures. They can also apparently use past experiences to guide current behavior, and even bacteria give similar behavioral evidence of learning and memory. Some have even argued that the fact bacteria can ultimately outwit antibiotics developed by arguably the most intelligent species on earth is evidence of a kind of primordial intelligence (but in this case, it is natural selection that results in antibiotic resistance, not intelligence).

Neither bacteria nor protozoa possess a nervous system, and almost no one (except for perhaps Deepak Chopra) would argue that they are conscious. However, because our social interactions are based on ascribing reasoning to the behavior of our fellow humans, it is common to conflate behavior with conscious intent.

As I alluded to before, we assume that the behavioral and psychological responses that occur in connection with conscious feelings are generated by those feelings. For example, we assume that we run away from danger (like the rat freezing when it sees a cat) because we are afraid. But what if we are afraid because we run? Or, what if running and being afraid are independent cognitive responses?

Confusion between consciousness and behavior can arise even among neuroscientists. In humans, the amygdala, a part of the brain located in the limbic system near the base of the brain, is often called the "fear center"—the part of the brain most closely associated with the emotions of fear and anxiety. Joseph LeDoux was one of the neuroscientists who helped reinforce this cultural meme, but he now says that was a mistake.

The amygdala is an essential part of the brain circuitry that controls behavior and psychological responses to threats. It is natural to conclude that it is responsible for generating the feeling of fear, but LeDoux now argues that the conscious feeling of fear arises from different cognitive circuits that are related to consciousness itself.

When flies sense danger, they stop moving. They lack an amygdala, but they have other cognitive survival circuits that detect and control responses to threats. The genes underlying such behavior are similar to those in mammals, which they may have inherited from some common ancestor hundreds of millions of years ago.

The flies may appear afraid, but they likely lack the circuitry that results in consciousness. We too have instinctual circuits that control behaviors associated with certain emotions that are likely based on ancient survival circuits. They don't make the emotions, however.

The evidence for this differentiation comes, as LeDoux has emphasized, from the failure of so-called anti-anxiety medications to actually treat the circuits that generate anxiety and fear. The medications may have changed the behavioral responses to threats in animals, including shortness of breath, muscle tensions, and heightened alertness, but in doing so, they would have targeted survival circuits. The medications might therefore help govern behavioral symptoms associated with anxiety in humans—symptoms that may need to be treated—but they do not target the conscious feelings of anxiety.

Returning to evolution, survival in organisms depends not merely on avoiding external threats, but also on sustaining metabolism within the organism by regulating its internal physical and chemical environment, including body temperature, waste disposal, energy generation, fluid intake, and so on.

While all biological organisms have methods of preserving homeostasis, as organisms become more complex, the requirements for homeostasis increase, and a more complex internal sensing mechanism is required. Precursor systems still exist in single-celled protozoan descendants called choanoflagellates, which form clonal communities. In these communities, there are molecular bridges between cells where cell communication occurs, and electrical signals are used to communicate within cells. All of this anticipates the basic molecular foundation of neurons, and choanoflagellates contain genes and proteins that animals would later utilize to form synapses, a key part of neuronal communication.

Much later, from an evolutionary perspective, a true central nervous system developed in animals, which served two purposes: to monitor external conditions and to monitor the internal state of the organism throughout its body. The central nervous system allows a high level of functional coordination between metabolic systems, aiding in maintaining homeostasis. In particular, it takes sensory information and generates motor responses.

Neurons have long fibers that extend out of their cell bodies. They generally have one long axon, used to send messages over long distances to other neurons, and many small fibers called dendrites that receive and transmit messages to and from nearby axons. The use of electrical signals rather than chemicals allows rapid communication across the nervous system so that some distance can be maintained between sensory cells and the cells responsible for motor responses.

Nerve cells facilitate immediate, innate threat responses, thus extending the chemical response mechanism in simpler species, a key development that changed the way organisms could behave and also paved the way for consciousness involving a novel aspect of neutrons. They can be modified when an organism interacts with the environment. This "synaptic plasticity" is perhaps the most important development that changed life for animals, because it allowed organisms to overcome what would otherwise be simple innate behavioral responses and instead produce a wide variety of responses. It forms the basis of learning in animals, and, as we shall see, it carries a key characteristic central to two pillars of modern human cognition: language and consciousness.

As the nervous system developed, along with increased complexity and capability in organisms, the central unit coordinating the whole show had to increase in complexity and capability. So, the need for a brain. With the need for a temperature regulation system, a motor response to predators, and increased visual acuity, and eventually for solutions to higher-level issues related to reproduction and the survival of offspring, brains had to evolve and increase in size, adding new functions and ultimately, enhanced cognitive circuitry.

The modern human brain has a complexity that continues to surprise us, even today, after several centuries of detailed anatomical findings. Most striking is the fact that the separate components—forebrain, midbrain, hindbrain, spinal cord, and so on—have functionalities that are not independent of the other components. For example, returning to the question of fear, it used to be viewed that that paleocortex encompassing the limbic system was responsible for primeval emotions like fear and aggression while the neocortex was responsible for higher order cognition but not emotions. The limbic system controls behaviors associated with aggression and defense, but the feelings associated with these behaviors are not necessarily causally the source of control of these behaviors. Moreover, the limbic system includes areas like the hippocampus and cingulate cortex that contribute to cognitive functions, including memory, while the neocortex contains areas that relate to emotional experiences.

The fundamental problem with understanding the nature and origin of consciousness is not so much one of isolating the separate cognitive brain functions that may contribute to the end result, but that the end result appears to be so much more than the sum of its separate parts. Awareness, feelings, memory, and learning all play a role, but even awareness of surroundings is very different from awareness of the self as it experiences the surroundings. The memory of events or dangerous situations is not the same as the memory of being in those situations and experiencing the emotions associated with those events. And so on.

And even emotions themselves vary in nature. Pleasure and pain seem primeval, as they are based on direct responses to external stimuli, but emotions like sadness, longing, anticipation, and distrust seem to belong to a higher order. And when trying to get a handle on the latter, it is difficult to try and use evolutionary arguments because all we can do in animals is observe behavioral consequences to external stimuli, probing perhaps the primal sensations of pleasure and pain, but gaining no insight into any possibility of cognitive introspection that could follow. My wife tells me my dog appears to be sad when I am away from the house. As touching as this sounds, it may reflect more her conscious concerns for me and my dog than explicit evidence that my dog has feelings.

In spite of the need to consider emergent properties of consciousness that go beyond exploring the functioning of specific components, one way to begin to explore the possible uniqueness of human consciousness is to examine those evolutionary anatomical brain developments that are unique to humans. As Joseph LeDoux has emphasized, the natural place to focus is the prefrontal cortex, which is most often viewed as the central command for cognition. It is the area of the cortex that differs most between primates and other mammals, and in which there are also microscopic differences between humans and other primates at the cellular, molecular, and genetic level. It is connected to all areas involved with higher order processing, including perception, memory, and language. And even here, there is a hierarchy of apparent processing, with the most anterior region, called the frontal pole, receiving inputs from many different cognitive zones and appearing to be involved in abstract conceptual reasoning, including planning, problem solving, and controlling deliberative behaviors. However, it is important to once again emphasize that cognition is not exclusively carried out in the prefrontal regions. There is constant communication with and feedback from other posterior regions, and significant cognitive abilities can remain even if the prefrontal cortex is damaged.

Size may not matter, but the prefrontal cortex was found to be relatively larger in humans than in other primates, although precise measurements suggest that this might be accounted for based on body size differences after all. What seems more relevant are differences in the cellular structures in our brains. Neocortical tissue in the prefrontal cortex and frontal pole has six different layers, with one layer having a unique cell type called granule cells. Only primates have granule cells in this layer of the prefrontal cortex, which suggest unique kinds of processing.

In addition, the human prefrontal cortex has a different spatial arrangement of cells, different levels of connectivity between cell layers, and different patterns of gene expression related to metabolism and synapse formation. Also, neurons in the prefrontal cortex in humans are more interconnected with neurons in other brain areas.

In this regard, Antonio Damasio has emphasized what he feels is a very important facet of consciousness (something we will come back to): the connection between the brain and the interior of the body as a source of conscious feelings. At a very basic level, a sense of self comes from being aware of one's body and the condition it is in. Feelings are not detached from the body structures that feel. To support this connection, there appear to be distinct relevant physiological differences in neurons involved in transporting signals from the body into the nervous system, the so-called interoceptive system.

As I have described, neurons have a cell body and an axon, the long cable that allows signaling to other distant neurons through a synaptic connection. The axon, being like a kind of electrical wire, is insulated by something called myelin, which prevents extraneous contact with the environment. If myelin is absent, molecules surrounding axons can alter its electronic firing capabilities. Also, in this case other neurons, beyond those with direct connections to the original axon at its synapse, can make contact with the axon along its body, resulting in what is called non-synaptic signaling. Myelinated axons are, by contrast, insulated from the influence of their surrounding environment. The majority of neurons involved in interoception, however, lack myelin sheaths, which makes them far more sensitive to their environment.

Another opportunity for interoception to comingle neural signaling with direct input from the body arises due to the lack of the traditional brain-blood barrier in regions of the brain associated with interoception, including in the spine and brain stem. The lack of such a barrier allows chemical signaling within the body to interact directly with neural signaling.

Recall that the evolution of the nervous system allowed improved homeostatic regulation of an organism through careful sensing and central processing, resulting in improved responses to threats or opportunities. If, as Damasio argues, "feelings" are the first step toward conscious self-awareness, these physiological features of interoceptive neurons are consistent with the nervous system, allowing "homeostatic feelings" to develop, so that the previously innate response to recoil from possibly painful stimuli could evolve into "feeling" pain and ultimately into selecting from a variety of possible conscious responses based on inference and reason. As Damasio stresses, feelings are among the first examples of phenomena of "mind." They represent not just a physiological response to external stimuli, but an introspective reflection on the state of our own bodies. In that sense, they represent a natural extension of the physiological process of homeostasis by which organisms regulate their internal states.

An additional physiological component that seems relevant to what eventually becomes consciousness is one of the early discoveries of cerebral processing. In the 1970s, John O'Keefe discovered that cells within the hippocampus actually create cognitive maps of the spatial environment around the observer based on various fixed orientation markers. For his discovery of these place cells, he shared the 2016 Nobel Prize.

The notion of cognitive maps reinforces the notion of consciousness as an intermediary between what would otherwise be a fixed innate connection between stimuli and responses, based on the creation of an internal representation of the external world. From that, an awareness that the organism experiencing the stimuli is physically present in the theater of operation arises, and from that, using memory and reasoning, one of many possible outcomes arise. As Susan Blackmore put it,

> *"The mind is like a private theatre inside my head, where I sit looking out through my eyes. But this is like a multi-sensational theatre with touches, smells, sounds, and emotions. And I can use my imagination to conjure up sights and sounds as though seen on a mental screen or heard by my inner ear. All these*

*thoughts and impressions are the 'contents of my conscious-
ness,' and 'I' am the audience of one who experiences them."*

The theater analogy, however, is of limited utility, not least because
much of what is going on in conscious cognitive processing is hidden,
even from the mind doing the processing. A famous experiment involving what is called "blindsight" involved a patient who had damage to
part of his visual cortex (V1) where the visual world is laid out in a
map. A small blind area was created where the patient could not see
anything. When the patient was presented with images of either vertical or horizontal stripes in that area, the patient said he could see no
stripes. But when asked to guess whether they were vertical or horizontal, the patient was correct nearly 90 percent of the time.

The result is understandable, as there are many parallel pathways
through which visual information is transmitted in the brain. Some
don't involve V1. But the key fact here is that even if this explains the
patient's ability to intuit the nature of the image he couldn't see, his
conscious awareness of the nature of that image was absent.

The parallelism in this visual processing is mimicked in many
other brain processes. Indeed, the brain appears to be a massively
parallel processing device. If the mind is a theater, it is a multiplex
with many movies going on at one time. Certainly, the presence of
many cortical regions where higher order brain processes appear to
be carried out implies there is not a central headquarters where consciousness itself resides.

How does life, then, appears to us as a single movie?

The brain is processing, at any one time, a massive amount of
information. Some neuroscientists liken it to a global workspace. I
envisage this workspace as being like the large screen on my computer,
which has many different windows on it, corresponding to many different processes that I have opened. I am currently staring at the word
processing window, but behind and to the right is my web browser.
Over further to the right is a notes window where I have written notes

and ideas relevant to this particular chapter. Down to the left is a mail window displaying current and past messages, and in the upper left are several windows listing files in various storage locations.

I am aware of the other windows, but my attention is focused on this word processor at the moment. But the other windows are still open.

There are now two ways of considering the role of consciousness in this picture. In the first, items in this workspace become conscious thoughts by virtue of being brought to the front layer of the monitor window—that is, being broadcast. In the second, all the windows are part of our consciousness.

The philosopher Dan Dennett has a slightly different take. He argues that things in the workspace are not really in or out of consciousness until we are probed in a certain way and respond to some stimuli. At that point, one part of the workspace is picked, and we decide we are conscious of that.

I find this vision of consciousness remarkably reminiscent of the classical picture of a quantum system, which is in many different states at one time. When we perform a measurement, we select one of the many alternatives. Until we make the selection, we cannot say the system is in any one state.

This is not to suggest that quantum mechanics has anything to do with consciousness, even though some, like the physicist Roger Penrose, who along with collaborators at the University of Arizona, has argued that quantum mechanics is an essential feature of human consciousness. I have listened to talks by members of this group, who, except for Roger, are not physicists, and frankly make little sense to me.

I brought up Dennett's idea not just because it is intriguing, but also because he is a philosopher. As I described earlier, the fact that debates around fundamental issues associated with how consciousness might become manifest are currently being pushed forward by philosophers as well as neuroscientists demonstrates what a nascent stage the science of consciousness currently finds itself in. We are not quite sure how to frame the right questions (something philosophy provides help with), much less know how to answer them. Even with the burgeoning knowledge of brain physiology and functioning, the fundamental question of how higher order cognitive processes associated with memory, learning, and reason lead us to be conscious of our own existence still remains one of the most elusive and fundamental questions science has yet to be able to resolve.

Another possible line of attack begins by considering the possible distinct evolutionary advantages that consciousness might endow humans with. Take the notion described earlier that feelings emerged as ever more complex systems evolved to incorporate higher order cognitive processing to issues of survival and homeostasis. Consciousness, through introspection, could build on the nervous system monitoring of basic internal body conditions to produce novel, rather than innate, survival strategies.

The ability to use internal representations of goals, whether from cognitive maps or stored memories, to flexibly respond to the changing environmental conditions, was a huge evolutionary leap, and has been noted to probably exist only in some mammals and perhaps in birds.

What additional developments then pushed hominids to a level where the recognition of those goals could be internalized, giving the organism a distinct sense of existence within that environment, allowing it to be both subject and object in decision-making?

One possibility is the evolutionary emergence of language. Here I think the viewpoint of Noam Chomsky is most compelling. While it is natural to think of the utility of language for the purpose of inter-organism communication, and there is little doubt that—from an evolutionary perspective—language gave early hominid social groupings a survival advantage, Chomsky has argued that language evolved as a part of neural circuitry that effectively generates thoughts, making modern cognition, reflection, and self-awareness possible.

Internally generated thoughts can then sometimes be externalized via a sensory-motor medium and then used for communication.

Possessing language as we do, it is virtually impossible to imagine constructing thoughts without the generative processes of language. One doesn't know what one's thoughts are until they are formulated by language.

Chomsky is not alone in emphasizing the generative connections between language and thought. Indeed, until the rise of behaviorism, with its emphasis on external inputs and outputs, the standard view was that language and thought were essentially indistinguishable.

The distinction between the mechanism of the generation of language and its later articulation as a form of communication is also so subtle that we find it difficult to find the appropriate language to describe it. The neurologist and writer Oliver Sacks wrote, "It is through inner speech that the child develops his own concepts and meaning; it is through inner speech that he achieves his own identity." It is clear that he is alluding to the importance of language in cognition, but is "inner speech" really the appropriate way of describing it? What the mind is producing in the process of generating language is more than simply a version of external speech with the articulation inactivated.

Psychologist Steven Pinker has stressed the fundamental connection between language and our very humanity, saying, "Language is a window into human nature." Finally, Dan Dennett has argued that, "The kind of mind you get when you add language to it is so different from the kind of mind you can have without language that calling them both minds is a mistake." Indeed, one can frame this perspective succinctly as follows: without language, there literally is no mind, just a brain.

What I find particularly compelling about this viewpoint—especially in light of the discussion of consciousness I have given thus far—is the similarity between the gigantic logical leap that language provides and the behavioral leap that consciousness provides, namely, a seemingly infinite flexibility to generate new cognitive states and responses to stimuli.

The formulation of new strings of words to make sentences that literally may never have been spoken or thought before is, from a computational perspective, a major cognitive leap—perhaps the most important such leap ever made, as Chomsky and other cognitive scientists like Steven Pinker have emphasized. It provides a window into detailed quantitative logical mental algorithms that may be hardwired in our brains.

Consciousness provides a similar behavioral leap. By understanding the self as both subject and object—with an internal representation of one's own bodily states and the state of the surrounding environment as well as an ability to forecast goals and internal and external consequences of those goals—consciousness provides us with a seemingly infinite choice of behavioral responses associated with each moment of our lives.

Some like to think of these choices in the context of free will, but I believe that opens up an unnecessary, and not particularly useful, can of worms. Whether or not we have free will, or simply the appearance of free will (which arguments from the perspective of physics would suggest) isn't really the important issue. It is that seven billion different humans may take actions that appear to reflect seven billion different life decisions when faced with similar external environmental pressures.

* * *

As we move from the realm of experimental physiology to theoretical neuroscientific and even philosophical speculation, which is where we currently sit when it comes to trying to capture the nature of consciousness and its origins, we inevitably face the elephant in the room: the concept of self.

It is claimed, certainly by Buddhists, and also by numerous cognitive scientists, that the self is an illusion. Certainly, everything we know about the brain suggests there is no wizard deep behind the curtain of our minds pulling the strings whom we can label as a kind of primordial "I" inside each of us. The brain is at the very least a distributed processing system, with multiple routes for information flow and processing, multiple centers, and many inputs for and alterations of cognitive states that we are not consciously aware of.

In 1985, neuroscientist Benjamin Libet described the results of an experiment that rocked his field. He asked subjects to watch a dot orbiting around a clock face on a TV screen and then note the position of the dot at the moment they randomly decided to flex their fist. At the same time, using electrodes on their scalps, Libet measured a gradually increasing signal, called the readiness potential, which signaled the start of brain activity leading up to subjects' action. He found the conscious decision to act came about two hundred milliseconds before the action, but that the readiness potential signal began around 350 milliseconds earlier than that, almost a half a second before the action, and over a third of a second before the subjects consciously decided to act.

While many scientists and philosophers debated the philosophical implications of this result, operationally the implication is clear. Consciousness is indeed like an iceberg, and we indeed experience only the small part above the water. There is a massive amount of cerebral processing going on behind the scenes. (Just how much remains to be seen. It is worth noting that at least one study, by Shaun Hegarty, argues that much or all of Libet's delayed result could effectively disappear if the speed of neural transmission is taken into account.)

James' "stream of consciousness" may be the perceived end result of our conscious awareness, but many neuroscientists view the persistent "I," who is the featured star in the movie of our lives, as a kluge. The mind cobbles together images, perceptions, rationales—sometimes after the fact—to produce what we perceive as the single coherent narrative we describe as self-awareness. The brain does what it does, without any external or internal observer who watches or controls what is going on.

Susan Blackmore suggests that this idea of self as an illusion goes all the way back to Hume, who, she describes, concluded, "The self is not an entity but more like a 'bundle of sensations'; one's life is a series of impressions that seem to belong to one person but are really just tied together by memory and other relationships."

One can take this notion further. If the narrative we experience is as much an invention of our minds as it is a true reflection of reality, one could argue that not just the self, but reality itself is an illusion. This is perhaps a more dramatic way of expressing a priori reasonable expectation that experience conforms to our modes of cognition. Neuroscientists Andy Clark and Anil Seth argue that because the narrative we cobble together is based on predictions generated from memory experiences as much as it is from actual external inputs, conscious perception is a "controlled hallucination." Of course, it is then worth recalling the important statement of Oliver Sacks, who wrote, in his book *Hallucinations*, that for those who experience them, the hallucinations *are* real.

What are we to make of all of this? Must we accept, as Blackmore argues, that this requires "completely throwing out any idea that you are an entity who has consciousness and free will, or who lives the life of this particular body"? That we should instead "accept that the word 'self,' useful as it is, refers to nothing that is real or persisting; it is just an idea or a word"?

I don't find this perspective either compelling or useful, any more useful than the claim, going back to the beginning of this book, that time may be an illusion. Once again, tell that to someone who just misses the train that would have taken them to an important job interview. As for the illusion of self, tell that to someone who has just been kicked in the shins or, worse still, been jilted by their lover.

Free will might be an illusion too, and indeed, some interpreted Libet's results as the final nail on the coffin of free will, as he suggested that decisions may be made in the brain well before we think we make them. But we live in a world which, for all intents and purposes, is indistinguishable from one in which it isn't. So, acting as if we have free will makes operational sense. Similarly, consciousness may have invented the illusion of self, but if consciousness governs who we are, then trying to rid the self from one's mind seems to me to be little more than a sterile exercise, unlikely to bring on any profound revelations about in the real world, about which, after all, our consciousness provides an entrée.

In short, we have to play our hand with the cards we are dealt. A scientific approach to human cognition should account for the fact that, operationally, as far as internal mental states are concerned, awareness of self is as real as awareness of the external world, and if we want to understand the latter, then debating the existential reality of the former may not be productive. As far as our minds are concerned, both "self" and "real" are inextricably linked by consciousness. What matters is how consciousness creates our experience of both. And it is that thorny and complex fundamental issue, as fundamental as any question we have about our place in the universe, which continues to challenge us.

* * *

"Neuroscience is telling us that our sense of self is an outcome of complex interactions between brain and body, of neural processes that update the self moment by moment, the moments strung together to give us a seamless feeling of personhood. We often hear of how the self is an illusion, that it is nature's most sophisticated sleight of hand. But all this talk of tricks and illusions obfuscates a basic truth: remove the self and there is no "I" on whom a trick is being played, no one who is the subject of an illusion."

—ANIL ANANTHASWAMY, *The Man Who Wasn't There*

Sometimes, if you really want to understand how some machine works, you need to find a broken one and try to fix it. And so, while the hand nature has dealt us limits our abilities to scientifically investigate consciousness in ourselves and others, sometimes nature screws up and creates a broken machine. This can give us a back door, or perhaps rather a rear window, into facets of consciousness that would otherwise be impossible to explore.

The disorder that inspired the title of Anil Ananthaswamy's book is called Cotard's syndrome, in which patients can believe they are dead, that the "I" inside their heads no longer exists, and therefore that they no longer exist. Like many such bizarre disorders discussed by Ananthaswamy, or by the neurologist Oliver Sacks in his fascinating books and articles, it is hard to believe that anyone can really suffer such delusions, but they do. By studying such patients who have lost their sense of self in a variety of ways, we can hope to learn something about the processes that make us feel like we do exist.

These illnesses include not just obscure and rare disorders like Cotard's, but ailments such as schizophrenia, which can cause individuals to feel like they are not themselves or are not in control of their own actions; Alzheimer's, in which patients eventually lose their personalities, memories, and fundamental humanity; and autism, which inhibits one's ability to "read" others' minds at the level that most of us instinctively do in the process of socially interaction.

For example, in the case of Cotard's, brain scans reveal significantly depressed activity in a variety of areas associated with the frontal lobe and the parietal lobe behind it that suggest this network is related to conscious awareness, with two sub-areas relating to awareness of the external world through the senses and one to the internal state of the body, including mental states. In addition, there is evidence from these studies that longer-distance communication between this frontoparietal network and another brain region, the thalamus, contributes to functions ranging from the promotion of arousal in response to stimuli to conscious awareness. It was suggested that the depressed metabolic activity in these regions in a Cotard's patient could be related to their reduced sense of self. In another Cotard's patient, there was atrophy in the frontotemporal region of the brain, particularly in a deep region called the insula, which is thought to contribute to the subjective perception of body states—which clearly is a key component of any sense of self.

In the case of autism, there is also work that suggests that the inability to relate to others is tied to parts of the brain that may contribute to awareness of one's own body and also its interactions with the environment. It is speculated that this produces a less certain sense of self, which then contributes to the observed autistic behavioral problems.

Illnesses like these may be tragedies for those involved, but they are backhanded gifts from an otherwise uncaring universe to scientists studying consciousness. By comparing the neural activity of those whose sense of self is severely altered, in one of a variety of ways, with brain scans of others without these illnesses, researchers have additional tools to hone in on those cerebral circuits that contribute to the processes these individuals have tragically lost. These won't directly reveal the actual mechanisms that result in conscious awareness and a sense of self, but they can provide useful empirical data that we would not otherwise have, had nature been more forgiving.

Indeed, maybe expecting actual physical mechanisms underlying consciousness to be revealed is asking for too much. Noam Chomsky has emphasized to me that trying to solve the "hard problem" of consciousness—of trying to unravel what it "feels like" to see the sunset—may be misplaced. He notes that in the seventeenth century, instead of the hard problem of consciousness, the philosopher and scientist William Petty referred to the "hard rock" in science: the problem of finding a way to mechanistically explain motion by way of a physical machine that one might construct, at least in principle. Eventually this rock was circumvented, as Newton and others found the world could be described directly through mathematical equations. But even Newton and his contemporaries felt that the mathematics alone was not satisfactory.

A similar example comes to mind related to ideas I have already discussed in another context. Maxwell, whose mathematical theory of electromagnetism I have described and extolled, nevertheless continued to present a mechanistic representation from which his equations could result, using mechanical wheels and pulleys. That representation fell by the wayside a long time ago and we now accept Maxwell's underlying mathematical equations as sufficient to explain all the phenomena of electromagnetism.

After four hundred years, modern physicists have come to accept Newton's and Maxwell's mathematics and Maxwell's in particular as reflecting the closest understanding of the physical world we can expect to have. Perhaps in the end, the "hard problem" of what it is like to smell a flower will be put aside with the recognition that the best we can hope for is that our empirical knowledge will allow us to develop a mathematical explanatory theory of consciousness without a representation via specific physical mechanisms.

Some time ago, the university institute that I directed ran a workshop entitled "The Origins and Future of Pattern Processing and Intelligence: From Brains to Machines." The idea was to bring together neuroscientists and computer scientists to see what they could learn from each other, with the aim of improving computer learning, but to also see what computer experiments might reveal about possible neural processes. One of the issues of interest to the computer scientists was the relationship of creativity to madness, so that computer scientists might incorporate lessons learned from cases of known damage to neural processing, to ensure that down the line, their eventual robots might not suffer from something like schizophrenia. (We organized a related public dialogue on this topic where I sat down with Johnny Depp, a brilliant actor who both in his personal and professional life has sometimes had to reconcile skirting the boundary between these two extremes of creativity and madness. You can still watch that event online.)

There is only so much we can learn from broken systems, however. The physicist Richard Feynman once said to me: "He who can do nothing, knows nothing." What he meant by this is explained in a more famous quotation found written on his blackboard after he died: "What I cannot create, I do not understand." In the context of this discussion, this implies that while having access to broken machines can be helpful, if we cannot build something from scratch ourselves, from raw components, without a blueprint from others, then we really don't understand how it works. When it comes to consciousness, if indeed we are prisoners of our own consciousness, perhaps the one way we will really be able to ensure we understand its origins is to build a conscious machine from scratch.

There are mixed ideas about whether this will even be possible, and what will happen to the world if we can. Most of these arise from trying to extrapolate current efforts at what has become known as artificial intelligence (AI) technology. I have never liked the term artificial intelligence as it applies to the technologies we are creating, because there is, as far as I can see, nothing artificial or intelligent about it.

Most of what is now classified as AI is really a case of machine learning (ML). With ML, computers sift through mountains of data. Faced with increasing amounts of data, real and imagined, internet companies are under immediate pressure to help develop software and hardware that can "learn" from the data—that is, systems that can adapt to future real-time data inputs, using information gleaned from scouring massive amounts of past data.

Fortunes, both economic and political, have risen and fallen based on the ability to exploit vast user demographics to direct advertising to the right place at the right time or generate false news stories and distribute them to where they might most effectively influence voting patterns.

But perhaps the most well-known example of this trend involves the effort to develop driverless cars, so that they can distinguish stop signs from pedestrians and bicycles from trees. The efforts so far have been mixed, with some cars appearing to drive intelligently, but also with numerous tragic examples of false deductions.

Most of these developments involve an algorithmic methodology called neural networks. These are modeled on what early researchers thought modeled learning in the brain. Basically, these networks "learn" by trial and error. Many different connection pathways are explored, each of which can lead to different results. Connection algorithms can then be altered and optimized in response, with those pathways leading to the desired result, getting weighed more strongly in each successive run.

These software and hardware developments are occurring at a pace far exceeding our ability to coordinate them—partly because Moore's Law has persisted well past its expiry date, but also because of new hardware. As a result, neural networks are beginning to handle specific and limited complex tasks much better than the human brain does, with improved efficiency and significant reduction in energy costs. Using these methods, computers can learn to play Go better than any human player, and can analyze some images, like X-rays, better than many doctors. The results seem like magic, and of course raise concerns among many that as these systems exceed human capabilities, human control will also go by the wayside.

These machines are essentially black boxes, with a challenge or question posed at one end and a result coming out the other. Because there is no explicit logical analysis explaining the process, even if the results are accurate, they can be uncomfortable for humans, who like to understand the underlying reasoning in order to accept the results. For example, say a neural network medical diagnostic machine took as input the results of many medical tests you have had and then prescribed a course of action. Would you be willing to follow the course if no one, including doctors who might be coordinating the treatment, knew why it might work?

Regardless of how remarkable the abilities of these new devices may seem, it is not at all clear that they are doing anything close to "thinking," at least as we imagine it is happening in the human brain. While the machines have an ability to sift through orders of magnitude more data in a short time than humans can, the actual processing of the data, which for the machines is really just sifting data, may be dwarfed by the processing capabilities of the human mind.

To get a sense of the challenge of creating a thinking, self-aware machine based on current forms of electronic computing, consider energy issues. Several years ago, I read an estimate that to match the storage and processing capabilities of the human mind in a way that was necessary (but perhaps not sufficient) to produce consciousness would probably require something like ten terawatts of power. I expect today the number could easily be several orders of magnitude smaller. (Indeed, just after I wrote this, I learned of a new development, involving so-called "neuromorphic chips" that mimic short term memory storage in the brain, which could reduce energy consumption in AI algorithms by perhaps three orders of magnitude.) No matter. The human brain uses about twenty watts, less than the power used by the laptop I am using to write this on (and even though my Mac may be smarter than a Windows-based machine, neither is approaching self-awareness). The difference between these two quoted numbers is a million million. So even if today the actual comparison is only a thousand million difference in energy consumption, it is clear that the brain is currently doing something quite different than my computer is.

It may be that some aspect of neural networks does mimic actual neural processing associated with human decision making. While it really doesn't "feel" like the kind of thinking we are conscious of when we make decisions, I have described how it is also clear that a lot of our decision-making happens under the radar. Regardless, no one is suggesting these devices are in any way "aware" of what they are doing.

Evolutionary arguments, like those put forward by Antonio Damasio, Joseph LeDoux, and others suggest that devices that follow this approach, no matter how fast or complex they become, are unlikely to become conscious. A key facet of our consciousness arises from our ability to sample and sense our own internal states. In the language of Damasio, it is our feelings that count. These feelings derive from the homeostatic demands of organisms with complex nervous systems. They were a necessary precursor to human intelligence. He has argued that future robots, if they are to have any hope of self-awareness, will need to have a "body" that requires regulation in order to persist. As he has put it, "What the machine 'feels' in its body will have a say on the matter of responding to the conditions that surround it. That 'say' is meant to improve the quality and efficiency of the response."

Beyond feelings, our brain reflects four billion years of organismal evolution on this planet. It may be leaps and bounds beyond the neural capabilities of many other mammals, but it shares an evolutionary architecture with them. The evolution of hominids did not reinvent the neural wheel. It added some spokes, and changed the tread on the tires, and maybe improved the drive mechanism.

As LeDoux has emphasized, human emotions are possible because of the unique capacities of our brains, which depend on evolutionary developments for our early or more recent hominid ancestors, including the evolution of language, reasoning, and introspective processing. The neural circuits that resulted had to evolve on top of a framework that had already largely been established by billions of years of evolution, based on survival behaviors. The developments weren't trivial by any means, but their form is contingent on the initial framework. As LeDoux put it:

"Gaining an understanding of the animal heritage of nonconscious functions of the human brain is thus not a consolation prize. It is crucial for our understanding of both animal and

195

human behavior.... The universal strategies that survival circuits and behaviors tactically implement connect us to the entire history of life.... Only by knowing the whole story can we truly understand who we are, and how we came to be this way."

The key question, of course, is whether such an evolutionary history is also necessary to recreate consciousness in machines. Will devices that do not have the hierarchical framework of our brains—with forebrain and midbrain built on hindbrain, with succeeding layers taking on new tasks while interfacing and communicating across the entire system, all the while being intimately tied to bodily sensory information flow and regulation—be able to achieve consciousness?

Time will tell. It may be that fundamentally new computational devices, perhaps based on quantum computing ideas, will be required. I personally suspect, no matter how challenging, that there is no fundamental roadblock getting in the way of ultimately developing functioning self-aware machines.

For many, this thought is truly terrifying. There are already conferences on how to impose something like Asimov's Three Rules of Robotics on computing algorithms for AI so that these systems don't get out of control and destroy the world, like the robots of *Terminator* movies. I have attended talks by distinguished philosophers that described the need to input "universal human values" into AI algorithms. I found and continue to find that idea somewhat implausible at best. I am not sure what "universal human values" are, but one thing I am certain of is that machines will not learn them by the current process of sifting through existing data on human actions throughout history. It may be a matter of "do as I say, not as I do" when it comes to programming such devices, but that then begs the question of who will be inputting the "do."

One can imagine many scenarios in which things go south, and science fiction writers, futurists, and philosophers have run rampant in this regard. I take solace in several things.

First, science fiction trends miss the most interesting human developments. When I was young, I was led to believe by science fiction writers and futurists alike that by now I would be living in a world of flying cars and space tourism. Instead, I'm living in a world that no one envisaged, dominated at every level by the internet, which has changed both communication and information processing more radically than almost any other development in human history. There's a fundamental difference between science fiction and scientific discovery. The former extrapolates a future based on the present, whereas the latter creates a new present.

Second, I view all claims that we are on the verge of a singularity, where self-aware self-programmable computing machines take off, leaving us in the dust, as unrealistic. This is not just around the corner. Computers can beat us at playing Go, but I am told that many robotic systems still have problems folding laundry. And even the most successful deep machine learning programs may not have any relationship to the way actual conscious systems work in the real world. The question of whether neural nets are adequate to handle the kind of computability needed for cognition remains debatable. It is one of the motivations of Penrose and others for thinking about different kinds of computation at the cellular level.

Feynman himself speculated about quantum computing not just because of its possible utility in computation, but, at least in part, because he was interested in what quantum computers could reveal to us about quantum mechanics. Devices whose computations rely directly on the weird and wonderful features of the quantum world might reveal new insights on how to understand it better intuitively, without recourse to outmoded classical interpretations. I am fascinated by the possibility of finding out what another intelligence, perhaps one based on completely different sorts of neural modeling circuits, might have to teach us. What physics questions would such systems find fascinating? And what questions, what currently known unknowns, could they help us answer?

At the very least, the development of such machines would forever change our understanding of consciousness, potentially resolving one of humankind's most persistent ongoing mysteries.

Certainly, a world in which we might not be the dominant cognitive organisms would be a different world. But why must it necessarily be worse? Must a future in which machines can carry out many human tasks better than humans be awful? Would having more time to read, to explore the arts, to simply do all those things we like to do, while the engine of society is run by intelligent machines, automatically be such a bad future?

It is naïve, of course, to assume that the fruits of a world with enhanced productivity, with a potentially more sustainable infrastructure, would be shared evenly with all of humanity. Those who first develop truly autonomous thinking machines may have a tremendous economic advantage, and if history provides any examples, that might simply lead to further concentrations of wealth and resources in the hands of fewer and fewer people and institutions. But hey, one can always hope.

Fundamentally, the greatest change I expect may occur if self-aware machines are developed is that our new understanding of consciousness will change our understanding of what it means to be human. It may be necessary, in order to compete, for biological systems like humans to incorporate some of the machinery of these new technologies into their functioning and activities. New hybrid organisms may come into being. But they don't have to resemble the Borg.

A derivation of the Phoenician alphabet was introduced into ancient Greece in the ninth century BCE and evolved over the next few hundred years. This was viewed by some as the end of civilization—in particular, the end of literature and drama. Plato criticized written language as an impediment to wisdom, arguing that writing did not capture the whole truth and only propagated the *illusion* of knowledge. He also complained that writing would eliminate the need for memory. Socrates also viewed writing with suspicion, arguing that written communication would never be as clear as face-to-face communication because one couldn't ask questions of or argue with a document.

These complaints may seem anachronistic, but in the context of modern digital communication, they have a familiar ring. How many of us have groused that being able to google information will mean that people won't have to remember anything or work out details for themselves? Or that texting and tweeting are degrading human interactions, and in particular the ability to read and write?

What is called artificial intelligence is a natural by-product of human development, just as writing was then. Technology, a human invention, changes the world, but it also requires humans to change with it. This has been going on as long as humans have been human.

An AI future may be better than the present. After all, I imagine most people today think a world of books to read is better than the world was before writing allowed them to be written.

Recognizing that we don't know what the future will bring, just as recognizing how many questions about the universe are yet to be answered, may help us ensure that the future always remains more exciting than the past.

EPILOGUE

"*Dream or nightmare, we have to live our experience as it is, and we have to live it awake. We live in a world which is penetrated through and through by science and which is both whole and real. We cannot turn it into a game simply by taking sides.*"

—JACOB BRONOWSKI

Recognizing that we don't have all the answers is the beginning of science, just as it is the first step on the road to wisdom. That may seem trite, but it is worth repeating at the current time, as popular discourse appears more and more to be governed by ideological certainty than by rational inquiry.

If ideology intrudes on the process of science, the inevitable gaps in our knowledge can be exploited by those who wish to further their own ends, whether they're economic, political, or religious.

We are told that science will never explain love or that science cannot possibly present an accurate or complete picture of this portion of reality because it has been performed by individuals who themselves are flawed, for one reason or another—by their birth identity, their politics, or their position of wealth or stature—and that we need to root out or ignore their contributions if we are to achieve real progress.

These are games, pure and simple. It takes remarkable conceit to claim that science will never explain this or that because it implies that you know enough to know what we can never know. Equally, who is to decide who is or is not worthy enough to contribute to the process? Science proceeds by dialectic, where all ideas are subject to debate and attack, and bad ideas get rooted out precisely because the community as a whole has goals that transcend the specific penchants of individual scientists.

Motivated on the left by social justice concerns or on the right by conservative intransigency, pundits and politicians alike claim to know what is best for others as well as to know exactly what the causes of society's current ills are. They think they can somehow do this without asking skeptical questions or investigating the actual data.

Even universities, which should be the last bastion of skeptical questioning and open inquiry, are quickly becoming overburdened by issues of political correctness and claims of systemic ills that govern what can be said and who can say it. Sadly, the ability to question any statement, a sign of interest in learning and the hallmark of the scientific method, is too often stifled because concerns about offense, marginalization, and victimhood are becoming paramount.

My concern here is not university politics, however. I began this epilogue with a quote from one of my intellectual heroes, Jacob Bronowski, because I care about science and its ability to help us better understand nature and ourselves, to help create technologies that can improve our lives and our environment, and to better predict future alternatives.

Four hundred years of modern science have brought us to where we now stand, but where we go from here will not just depend on how we use our existing knowledge, but how we build new understanding of the world around us.

In this regard, recognizing our current limitations is an essential first step. Being able to quantify uncertainty, which is nothing other than explicitly understanding what you don't know so that you can determine the impact of that lack of knowledge on what we can say with confidence about nature, is probably science's greatest strength. As I have often said, not understanding something is not evidence for God or human frailty. It is just evidence of not understanding. And it should be an invitation to explore and learn.

Humility and honesty demand that we be clear about the limitations of our knowledge, but we shouldn't be shy about this. We should celebrate it. There remain remarkable mysteries to be uncovered.

As I remarked at the beginning of this book, focusing on the edge of knowledge has provided me an opportunity to explain how far we have come, but it also allows me to present some guideposts for the future. When I was a teenager, I remember reading a book by Richard Feynman called *The Character of Physical Law*. I had always been interested in science, but that book made it clear to me for the first time that the really interesting questions in physics had not all been answered yet. It was an invitation to me, and it presented a challenge and an opportunity to try and move on to the next step. I don't presume that this book will have the same impact on some young person today, but I hope it does. Indeed, that is why I have written it.

ABOUT THE AUTHOR

Lawrence M. Krauss is an internationally known theoretical physicist and the author of hundreds of scientific articles and numerous popular books including NYT bestsellers, *The Physics of Star Trek* and *A Universe from Nothing*. He received his PhD from MIT and then moved to the Harvard Society of Fellows. Following eight years as a professor at Yale University, he was appointed as a full professor with an endowed chair while still in his thirties. He has made significant contributions to our understanding of the origin and evolution of the Universe and has received numerous national and international awards for his research and writing. He is currently President of The Origins Project Foundation and host of The Origins Podcast with Lawrence Krauss. Between 2006 and 2018, he was Chair of the Board of Sponsors of the Bulletin of the Atomic Scientists. He appears regularly on radio, television, and film.

INDEX

INDEX